Manuel Konrad Huber

Modeling the Dynamics of Partial Wetting Using Smoothed Particle Hydrodynamics (SPH)

Logos Verlag Berlin

λογος

D 93 (Dissertation Universität Stuttgart)

Bibliografische Information der Deutschen Nationalbibliothek

Die Deutsche Nationalbibliothek verzeichnet diese Publikation in der
Deutschen Nationalbibliografie; detaillierte bibliografische Daten sind
im Internet über http://dnb.d-nb.de abrufbar.

ISBN 978-3-8325-4818-6

Logos Verlag Berlin GmbH
Comeniushof, Gubener Str. 47,
D-10243 Berlin

Tel.:   +49 (0)30 / 42 85 10 90
Fax: +49 (0)30 / 42 85 10 92
http://www.logos-verlag.de

Modeling the dynamics of partial wetting using Smoothed Particle Hydrodynamics (SPH)

Von der Fakultät 4 für Energie-, Verfahrens- und Biotechnik
der Universität Stuttgart zur Erlangung der Würde
eines Doktors der Ingenieurwissenschaften (Dr.-Ing.)
genehmigte Abhandlung

Vorgelegt von
Manuel Konrad Huber
geboren in Tübingen

Hauptberichter: Prof. Dr. Ulrich Nieken
Mitberichter: Prof. Dr. Seyed Majid Hassanizadeh

Tag der mündlichen Prüfung: 14.09.2018

Institut für Chemische Verfahrenstechnik
der Universität Stuttgart
2018

Vorwort

Die vorliegende Arbeit wurde im Rahmen des internationalen Graduiertenkollegs NUPUS und meiner wissenschaftlichen Tätigkeit am Institut für Chemische Verfahrenstechnik (ICVT) der Universität Stuttgart angefertigt.

Daher möchte ich mich zunächst besonders bei den Personen bedanken, die mir die Erstellung dieser Arbeit auf wissenschaftlicher Seite ermöglicht haben. Allen voran ist dies Prof. Dr. Ulrich Nieken, der als Doktorvater und Institutsleiter zum einen die Voraussetzungen für meine Tätigkeit am ICVT geschaffen hat. Zum anderen hat er mir aber auch durch seinen Beitrag eine persönliche Entwicklung ermöglicht, wodurch er maßgeblich am Erfolg dieser Arbeit beteiligt ist. Des Weiteren bedanke ich mich sehr herzlich bei Prof. Dr. Seyed Majid Hassanizadeh. Durch seine Betreuung habe ich stets die geeignete Hilfestellung und Motivation erfahren, welche mir heute noch in Verbindung mit dem aufgebrachten Engagement als Vorbild dient.

Ein großer Dank geht ebenso an Stefan Dwenger und Franz Keller. Ihre Vorarbeiten bildeten zusammen mit den unzähligen fachlichen Ratschlägen und Diskussionen eine wichtige Grundlage für diese Arbeit. Ebenso bedanke ich mich bei den Kollegen Philipp Günther, Jeremias Bickel, Christian Spengler, Carlos Tellaeche, Winfried Säckel, Jens Bernnat, Rene Kelling, Manuel Hopp–Hirschler und Philip Kunz sowie bei den Kolleginnen Karin Hauff und Simone Seher für die tolle Arbeitsatmosphäre und die außerordentlich gute Zusammenarbeit am Institut. Neben der kollegialen Kooperation haben aber auch Thorsten Woog, Daniel Dobesch und Vanessa Hägele im Rahmen ihrer studentischen Arbeiten sehr zum positiven Gelingen der Dissertation beigetragen.

An der Stelle sind auch Dr. Gheorghe Sorescu, Katrin Hungerbühler und Holger Aschenbrenner hervorzuheben. Mit ihrem Einsatz halten sie das Institut in den grundlegenden

Belangen fortwährend am Laufen und liefern somit die Basis für das alltägliche, erfolgreiche Arbeiten am Institut.

Ein besonderer Dank geht ebenso an Prof. Dr. Rainer Helmig. Er hat nicht nur das internationale Graduiertenkolleg NUPUS ins Leben gerufen, sondern mit seiner überaus motivierenden und offenen Art auch stets die Qualität von wissenschaftlichen Beiträgen auf internen wie externen Veranstaltungen gefördert.

Abschließend möchte ich mich vor allem auch bei meiner Familie, insbesondere bei meiner Mutter, für die jahrelange Unterstützung bedanken. Der positive Abschluss dieser Arbeit lässt sich nicht zuletzt auf ihr Engagement zurückführen.

Mönsheim, im November 2018   Manuel Konrad Huber

Contents

Nomenclature

Abbreviations

CA	Contact angle
CLF	Contact line force
CSF	Continuum surface force
DCA	Dynamic contact angle
ISPH	Incompressible SPH
MPS	Moving particle semi–implicit method
SCA	Static contact angle
SPH	Smoothed particle hydrodynamics
WCSPH	Weakly compressible SPH

Latin Letters

$\boldsymbol{a}$	Acceleration
A	Surface area
A	Arbitrary field quantity
A_I	Interpolated value of arbitrary field quantity A
$\bar{A}_I(r)$	Modified interpolation of arbitrary field quantity A
c	Color value/function
$\bar{c}$	Modified smooth color value
$[c]$	Jump in color value across interface
d	Diameter
d_c	Capillary diameter
d_s	Straitened diameter of nozzle
d_w	Outer diameter of diffuser
$\boldsymbol{d}$	Distance vector with respect to the wall
$||e||$	Error

E	Total energy	
f	Surface fraction	
$\boldsymbol{f}$	Force	
$\boldsymbol{F}$	Force (per volume basis)	
$\boldsymbol{g}$	Gravitational acceleration	
h	Smoothing length	here $h = 2.1\,l_0$
h_c	Half channel width	
$\boldsymbol{I}$	Identity matrix	
$\boldsymbol{J}$	Flux in control volume	
k	Spring constant	
l	Distance/spacing	
L	Length	
$\boldsymbol{L}$	Angular momentum	
$\boldsymbol{L}_a$	Correction matrix	
m	Mass	
$\boldsymbol{n}$	Normal vector	
$\hat{\boldsymbol{n}}$	Unit normal vector	
p	Pressure	
$\boldsymbol{P}$	Linear momentum	
Q	Adiabatic energy	
$\boldsymbol{r}$	Position/distance vector	
$\hat{\boldsymbol{r}}$	Unit vector of $\boldsymbol{r}$	
R	Radius of curvature	
t	Time	
T	Cycle time	
$\boldsymbol{T}$	Stress tensor	
u	Specific internal energy (per mass basis)	
U	Internal energy	
$\boldsymbol{v}$	Velocity vector	
V	Volume	
$\dot{V}$	Volumetric flow rate	
x	Position (1D)	
$\hat{x}$	Maximal amplitude	

| W | Kernel function |
| $\widetilde{W}$ | Shepard correction of the kernel |

Greek Letters

α	Contact angle	
δ	Volume reformulation	
Δt	Discrete time step size	
ε_a	Normalization	$\varepsilon_a = \sum_b V_b W(\boldsymbol{r}_{ab},h)$
η	Kinematic viscosity	$\eta = \mu/\rho$
κ	Curvature	
μ	Dynamic viscosity	
$\boldsymbol{\nu}_{ij}$	In–plane vectors of the ij–interface, perpendicular on the contact line, with $i,j = w,n,s$, cf. Fig. 3.6	
$\hat{\boldsymbol{\nu}}_{ij}$	Unit vectors of $\boldsymbol{\nu}_{ij}$	
ξ_{ab}		$\xi_{ab} = 8m_b \frac{\eta_a+\eta_b}{\rho_a+\rho_b} \frac{1}{|\boldsymbol{r}_{ab}|^2+\zeta^2}$
ρ	Density	
$\bar{\rho}$	Reduced fictitious density	
$\rho_k \psi_k$	Material quantity in control volume	
σ	Surface tension coefficient	
$\boldsymbol{\tau}$	Deviatoric stress tensor	
ϕ	Body source in control volume	
Φ	Arbitrary bump function	
Φ	Weighting function for surface tension boundary condition	
χ_{ab}		$\chi_{ab} = V_b \frac{4}{\rho_a+\rho_b} \frac{\boldsymbol{r}_{ab} \cdot \widetilde{\nabla}_a \widetilde{W}_{ab}}{|\boldsymbol{r}_{ab}|^2+\zeta^2}$
ω	Fitting quantity	

Superscripts

$'$	Indication for a continuous space quantity
$'$	Indication for an intermediate value
$*$	Heterogeneous surface
$\perp$	Normal component with respect to the wall

‖	Parallel component with respect to the wall
g	Ghost particle
i	Species i in a heterogeneous surface
net	Net contribution
t	Transpose of a tensor
vol	Volumetric description (per volume basis)

Subscripts

0	Initial value
I	1st part
II	2nd part
a	Particle a
ab	Particle a and b
act	Actual value
anis	Anisotropy
A	Advancing contact angle
b	Particle b
b	Bubble
c	Capillary
CL	Contact line
CS	Cubic spline kernel
D	Dynamic contact angle
f	Fluid phase
f	Fictitious value
g	Gaseous phase
G	Gaussian kernel
H	Hyperbolic kernel
int	Interpolation
int	Intermediate value
k	Phase k
l	Liquid phase
max	maximal value
n	Non–wetting phase

n	Number of discrete time step
o	Oil phase
quad	Quadrature
R	Receding contact angle
ref	For reference
s	Solid phase
S	Static contact angle
set	Set value
tot	Total value
w	Wetting phase
w	Water phase
W	Wendland kernel
wn	Interfacial quantity of wetting and non–wetting phase
wns	Contact Line quantity (wetting, non–wetting, solid phase)
x	x–component of a vector
y	y–component of a vector

Operators

∇	Nabla operator
$\tilde{\nabla}$	Gradient corrected Nabla operator
$\otimes$	Dyadic product

Dimensionless numbers

Bo	Bond number	$Bo = \frac{\rho_l g d_c^2}{\sigma}$
Ca	Capillary number	$Ca = \frac{\mu v}{\sigma}$
Fr	Froude number	$Fr = \frac{v_{in}^2}{d_c g}$
Ga	Galilei number	$Ga = \frac{\rho_l^2 d_c^3 g}{\mu_l^2}$
Re	Reynolds number	$Re = \frac{v_0 L}{\eta}$
We	Weber number	$We = \frac{v^2 d_c \rho_b}{\sigma}$

Zusammenfassung

In der vorliegenden Arbeit wird ein neues physikalisches und numerisches Modell zur dynamischen ortsaufgelösten Simulation von Zweiphasenströmungen unter Berücksichtigung von Oberflächenspannung vorgestellt. Besonderes Augenmerk gilt hierbei der Implementierung von Benetzungseigenschaften mit den daraus resultierenden statischen und dynamischen Kontaktwinkeln.

Zu Beginn der Arbeit wird zunächst die Grundlage für die numerische Beschreibung des Modells geliefert. In der numerischen Strömungsmechanik sind hierbei vor allem die Finite–Differenzen oder die Finite–Volumen Methode geläufig, wobei in der vorliegenden Arbeit in dem Zusammenhang ein eher moderner Ansatz verfolgt wird. Die verwendete Methode wird als "Smoothed Particle Hydrodynamics (SPH)" bezeichnet und bringt die Besonderheit mit sich, dass die Diskretisierung nicht wie bei herkömmlichen Methoden im Raum stattfindet, sondern in sogenannten Masse–Elementen, welche auch als Partikel bezeichnet werden. Diese Partikel sollten nicht mit molekularen Teilchen verwechselt werden, da sie vielmehr eine Ansammlung davon auf der Kontinuumsskala darstellen, für deren Beschreibung bereits die Gesetze der Kontinuumsmechanik gelten. In diesem Zusammenhang wird die Methode oft auch als partikelbasiertes Verfahren bezeichnet und zählt damit zu den Lagrange'schen Verfahren. Der Vorteil liegt darin, dass diese Partikel sich frei und kontinuierlich im Raum bewegen können. Das bedeutet, dass sowohl Ränder wie auch Strukturen im Rechenbereich nicht mehr an ein zugrunde liegendes Gitter gebunden sind und somit beispielsweise der Verlauf von gekrümmten Grenzflächen auch nicht mehr in diskreten Gitterschritten erfolgen muss. Folglich ist die Erwartung, einen glatteren Verlauf von Grenzflächen während der Simulation ohne aufwendige Gitterverfeinerung zu erhalten.

Um physikalische Größen an einem beliebigen Ort im Raum auswerten zu können, verwenden gitterfreie Verfahren wie SPH Interpolationsfunktionen, welche auf einer Faltung mit den sogenannten Kerneln basieren. Diese stellen eine im Raum verteilte Gewichtungsfunktion dar, mit deren Hilfe die gesuchte Größe am gewünschten Ort interpoliert wird. Somit kann nicht nur eine Größe wie zum Beispiel das Geschwindigkeitsfeld beschrieben werden, sondern auch dessen Ableitung. Bei der Transformation von der analytischen hin zur diskreten Schreibweise ist vor allem auf die Formulierung zu achten. In der Arbeit werden verschiedene Varianten aufgezeigt, die jedoch alle gewissen Besonderheiten unterliegen und nur in der richtigen Kombination gewünschte Eigenschaften wie Impuls–, Drehimpuls–, Masse– oder Energieerhaltung mit sich bringen. Im Zusammenhang mit der Zeitintegration wird in der Arbeit erstmalig ein Leap–Frog Verfahren für das inkompressible SPH eingeführt. Zum Abschluss der numerischen Grundlagen wird ein Überblick über die üblichen Formulierungen von Gradient– und Laplace–Operatoren samt Vor– und Nachteilen gegeben.

Im nächsten Abschnitt wird das physikalische Zweiphasenmodell Schritt für Schritt aufgebaut. Als Grundlage dienen hier die Navier–Stokes Gleichungen für inkompressible, Newton'sche Fluide. In Bezug auf die Beschreibung von Oberflächenspannung an Grenzflächen zwischen zwei Phasen wird das gebräuchliche "Continuum Surface Force (CSF)" Modell angewendet. In der Kontinuumstheorie werden an Grenzflächen regulär Randbedingungen formuliert. Durch eine Reformulierung der auftretenden Kräfte aus der Randbedingung werden diese auf ein kompaktes Volumen um die Grenzfläche neu verteilt. Um nun zusätzlich Benetzbarkeiten mit dynamischen Kontaktwinkeln beschreiben zu können, wird darüber hinaus eine neue Kontaktlinienkraft, die "Contact Line Force (CLF)", eingeführt. Dabei wird, in Anlehnung an das vorige Modell, die Kraft an der Kontaktlinie auf ein kompaktes Volumen um die Kontaktlinie neu verteilt. Das vollständige Zweiphasenmodell ergibt sich folglich aus einer Kombination von Navier–Stokes Gleichungen, Continuum Surface Force und der Contact Line Force.

Infolge der physikalischen Modellgleichungen und der numerischen Grundlagen wird im nächsten Schritt das vollständige numerische Modell samt Randbedingungen für glatte und gekrümmte Grenzflächen aufgezeigt und mit verschiedenen Testfällen validiert. Zur Bestätigung des Leap–Frog Zeitintegrationsverfahrens wird ein künstliches Fluidpendel eingeführt und eine oszillierende Bewegung aufgeprägt. Die zeitliche Entwicklung der Gesamtenergie beim Euler Verfahren weist ein deutlich divergierendes Verhalten auf, sodass die bessere Eignung des Leap–Frog Verfahrens verdeutlicht wird. Die korrekte Formulierung der örtlichen Randbedingungen wird für zwei verschiedene Viskositätsmodelle anhand einer Couette– sowie Poiseuille–Strömung für je eine Ein– bzw. Zweiphasenströmung dargestellt. Anhand einer Rayleigh–Taylor Instabilität wird anschließend die Interaktion zweier Phasen durch den Vergleich der Ergebnisse von SPH und OpenFOAM® validiert.

Nach dieser grundlegenden Überprüfung des Zweiphasenmodells wird anschließend auf die Oberflächenspannung eingegangen. Zunächst wird anhand des theoretischen Drucksprungs an einer gekrümmten Grenzfläche die erfolgreiche Implementierung der CSF Methode mithilfe der Young–Laplace Gleichung für den stationären Fall verifiziert. Des weiteren wird durch das Ausdehnen eines initial runden Tropfens entlang einer Hauptachse, verbunden mit der Kompression entlang der anderen, anschließend die Implementierung in Bezug auf die Dynamik überprüft. Nach den erfolgreichen Validierungsfällen der Zweiphasengrenzfläche wird im weiteren Verlauf der Fokus auf die Kontaktlinie gelegt. Die Fehler der Beschreibung bei statischen Kontaktwinkeln zwischen $30° - 140°$ liegt bei einer angemessenen Auflösung unterhalb von 5% und die Tendenz der dynamischen Kontaktwinkel ist in guter Übereinstimmungen mit den experimentellen Messwerten aus der Literatur. Zu guter Letzt bestätigt die "Gitterkonvergenz" eine erfolgreiche numerische Umsetzung der gesamten Oberflächenspannung.

Diese Testfälle bilden die Grundlage für die Anwendbarkeit des Modells in den verschiedensten Anwendungsgebieten. Exemplarisch wird hier auf zwei mögliche Bereiche genauer eingegangen. Zum einen lassen sich damit Gleichgewichtszustände von Benetzungen oder insgesamt der Zweiphasentransport durch ein Porennetzwerk berechnen. Dies kann z.B. genutzt werden, um ein allgemeines Systemverhalten zu erlernen, wodurch sich neue Geometrie– und Prozessoptimierungen ergeben können. Zum anderen gibt die Simulationsmethode Aufschluss über die Primärblasenbildung bei Blasensäulenreaktoren für unterschiedliche Materialeigenschaften unter Berücksichtigung verschieden geformter Dispergieröffnungen. Die Ergebnisse der Simulation und bereits publizierte empirische Modelle weisen dabei denselben Trend hinsichtlich ansteigender Volumenströme auf. Vor allem aber zeigt die Simulation im untersuchten Strömungsbereich, dass die verschiedenen Benetzungseigenschaften und Geometrien des Dispergierorgans einen Einfluss auf die Blasengröße haben und diese daher nicht weiter verallgemeinert und detaillierter untersucht werden sollten.

Abstract

In this thesis a new physico–numerical model for computational fluid dynamics with respect to two–phase flow is introduced. The special feature is the integration of surface tension with different wetting properties and the resulting static and dynamic contact angles.

First of all, the basic numerical description is presented. Common methods in computational fluid dynamics are for instance finite differences and finite volume schemes, respectively. However, in this thesis, "Smoothed Particle Hydrodynamics (SPH)" is used. The peculiarity of this method is, that the discretization is made by finite mass elements instead of a discretization in space. These mass elements are also called particles, but should not be confused with molecular particles, as they are rather an upscaled set thereof and thus follow the physical laws of the continuum scale. This way SPH is also referred to as particle–based method, which is assigned to the Lagrangian methods. The particles, which represent the discretization nodes, can move on continuous space. That is beneficial in terms of the description of curved boundaries and interfaces, because they are not related to an underlying grid structure anymore.

To be capable of evaluating physical field quantities at an arbitrary position in space, mesh–free methods like SPH introduce an interpolation based on a convolution with the so–called smoothing kernel. Thus, the average value of a desired quantity is evaluated at the desired position based on the values of the discretization nodes in the vicinity and based on the weighting by the smoothing kernel. This way not only quantities like velocity can be determined at any arbitrary position, but also their derivative.

Different numerical formulations for derivatives are presented in this thesis. In order to fulfill the conservation of mass, linear momentum, angular momentum and energy, beside the physical formulation, also the numerical description has to be adapted. With respect to the time integration, a Leap–Frog scheme is for the first time introduced in the context of incompressible SPH. An overview over common gradient and Laplace operators with advantages and disadvantages completes the description of the numerical basics.

With respect to two–phase flow, the physical model is built step by step. The basis is made of the incompressible Navier–Stokes equations for Newtonian fluids. To incorporate surface tension in the two–phase model, the "Continuum Surface Force (CSF)" is taken into account. When two phases are in contact, usually, boundary or interface conditions are formulated in conventional fluid dynamics. As the interface from a numerical point of view is not explicitly treated, interfacial forces are reformulated and redistributed to a small volume in the vicinity of the interface. In order to describe the contact with a third solid phase and to simulate related phenomena like wettability and contact angles, the "Contact Line Force (CLF)" model is introduced. The idea of the volume reformulation from the previous description is again utilized and adapted to distribute the forces of the momentum balance at the contact line to the particles in the vicinity. Hence, the whole two–phase model is a combination of Navier–Stokes equations, Continuum Surface Force and Contact Line Force.

With the knowledge of the two–phase model and the numerical basics, the complete numerical model, including the descriptions of boundary conditions, is built and validated. The Leap–Frog scheme is validated by an artificial fluid pendulum. The simulation reveals the divergent behavior of the total energy for the Euler scheme and this way confirms the advantage of the Leap–Frog scheme. The formulation of the boundary conditions is approved by a Couette and a Poiseuille flow experiment. Each is performed for a single–phase and a two–phase system in combination with two different viscosity models. The appropriate interaction of a two–phase system then is validated by the simulation of a Rayleigh–Taylor instability and matched with the results of OpenFoam®. After the successful validation of the two–phase flow without surface tension, the focus is now on the effects caused by the CSF. First the pressure jump at a homogeneous curved interface of a drop is approved and verified by the Young–Laplace equation. Then the drop is elongated in one direction and shrinked in the other direction. The resulting oscillation is compared for two different surface tension models to verify the dynamics. In order to prove the implementation of the CLF, static contact angles are simulated. The obtained error for contact angles in the range of $30° - 140°$ stays below 5% for an adequate resolution. Furthermore, the tendency of the dynamic contact angles is in good agreement with the experimental results from literature. Last but not least, the "grid convergence" confirms a successful numerical implementation of the whole surface tension model.

The previously mentioned test cases build the basis for the applicability in different areas. Exemplarily the focus in this work is on two different fields of application, where wettability plays a crucial role. For instance this model allows for the calculation of equilibrium states in porous sctructures. In this sense two–phase transport through a pore network is a very interesting area of application. The findings can be used to improve materials and processes by design. Another example is the simulation of primary bubble formation at bubble column reactors depending on material properties and different orifice types. The results show the same trend for increasing volume fluxes as empirical models do. But moreover, they reveal that bubble sizes differ, depending on wetting conditions and on orifice types. Therefore, the generalization of these conditions in the investigated flow regime is inappropriate, should be avoided and examined more in detail for future industrial applications.

1 Introduction

Surface tension has been an exceptional challenging phenomenon in science ever since the 19th century [Mar71]. In the early years the physical mechanism behind the phenomena was controversially discussed and investigations are mainly made by noting the phenomenological course of the process through observations with the human eye. Because the dynamics of these processes are quite fast and the resolution of the human eye is limited, Marangoni made his experiments in 1865 in the great basin of the Tuileries Garden in Paris, which has a diameter of 70 meters.

As a complement to the optical observations restricted by the human perception, few years later Lord Rayleigh made some physico–mathematical considerations on different types of surface tension experiments and published the first model equations for droplet dynamics in his work "On the Capillary Phenomena of Jets" [Ray79]. Under some restrictions these findings are still applied nowadays and used for test cases of droplet oscillations [Hu06, Woo11].

These early decades are characterized by the curiosity of scientists. In this way the ambition was just to understand the processes in nature and the fundamental observations were not driven by specific objectives for industrial applications. Later the benefit of surface tension including wettability was recognized and the perspective in research and development changed towards an output–driven purpose in different fields of application. The advantage of hydrophobicity for example then was revealed for the textile industry. This branch has a special interest to produce water–repellent fabrics in order to satisfy the demands of customers under unfavorable weather conditions [Wen36]. Beside the textile industry surface tension and wettability also play key roles in the chemical industry for products like washing agents, dish detergents and automotive paint, just to name a few of them. There surfactants are used to modify the wettability of an initial liquid by either being directly dissolved in water, where they act as solvent or degreaser and reduce the surface tension of water or they are applied on the counterside for instance by a coating, to obtain a water–repellent behavior. Not only surfactants can be utilized to change surface properties between two phases, but also the surface structure can be modified to obtain a desired wetting behavior as will be explained later. Moreover, the effects caused by surface tension are not only utilized for special functions of diverse products, they can also have a strong influence on production processes itself as soon as multiple phases are involved.

Beside the physical description of wetting, in this work a numerical implementation is additionally performed. Therewith the ability is generated to look at the dynamic wetting process for different surface structures under diverse experimental conditions.

Classical methods, like a Volume of Fluid (VOF) approach with an underlying Finite Volume Method (FVM), are stretched to their limits as soon as multiple phases with lots of advancing and changing interfaces are involved. Especially with respect to elaborate remeshing and dynamic adaptive grid refinement, which may become necessary under these conditions, the computational effort of this method dramatically increases.

During the last years particle–based methods come more and more into spotlight. The discretization is not made in space by a certain mesh but by finite mass elements, which can move continuously in space. This is why these methods are also called mesh–free methods. In this manner this discretization scheme is able to represent every shape in space, because the discretization nodes, the so–called particles, are not bounded by an underlying grid structure. Unfortunately there's no benefit for free. It comes at the cost of a in general higher computational effort. If the particles follow a Lagrangian motion, the relative particle arrangement among themselves changes and the particles change their neighbors over time. Therefore, neighbor lists have to be kept updated while for a fixed grid the cell arrangement is constant. But despite the higher computational effort compared to grid–based methods, particle–based methods are favorable in some applications. The latter are able to bridge the gap between continuum and fragmentation. An advantage with respect to astrophysical applications is, that no computation time is wasted for space, where no mass is present and therefore no impact is generated on the course of the simulation. Another advantage is the simulation of materials and fluids with complex shapes and interfaces as the particles can assume every point in space due to the Lagrangian description of the underlying physical process. With regard to multi–phase flow the introduction of further phases in an immiscible environment does not increase the computational effort, because it is not necessary to treat interfaces explicitly as long as interface phenomena are not taken into account. For grid–based methods the transition area always has to be defined on top of the regular numerical description. Smoothed Particle Hydrodynamics (SPH), the particle–based method applied in this work, inherently comes with a smooth description all over the domain and therefore, depending on the demands, no additional explicit treatment of interfaces is necessary. This is why for SPH in contrast to grid–based methods the computational effort does not dramatically increase when introducing further immiscible phases. Yet another advantage is the possible treatment of free surfaces, which results in a reduction of the computational effort by only considering the relevant dominating phase.

The aim of this work is to introduce a physical process description for the dynamics of partial wetting and further integrate it in the numerical framework of SPH in order to make the simulation approach capable of describing two–phase wetting dynamics for a variety of industrial applications.

The thesis is structured the following way. After this introduction first the basics of the numerical approach are explained in chapter 2. Subsequently the physical two–phase model is introduced in chapter 3. The knowledge of both of them is then applied in chapter 4 to build the complete numerical model followed by various verifications, which are presented in chapter 5. Last but not least applications of the new model are shown, e.g. the wetting of pores and primary bubble formation at bubble column reactors, see chapter 6.

2 Numerical approach: Smoothed Particle Hydrodynamics (SPH)

2.1 The interpolation

In contrast to grid–based methods, where the quantity always corresponds to a fixed point or cell in space, the idea behind a mesh–free discretization approach is the possibility to evaluate a physical field quantity $A = A(\boldsymbol{r})$ at every arbitrary position in space. According to Lucy [Luc77] and Gingold and Monaghan [Gin77] this concept is deduced from Monte Carlo theory, where a discrete set of particles is used to obtain a continuous representation of spatially varying state variables

$$A_I(\boldsymbol{r}) = \int\limits_{-\infty}^{\infty} A(\boldsymbol{r}')W(\boldsymbol{r} - \boldsymbol{r}',h)d^3r' . \tag{2.1}$$

Here $W(\boldsymbol{r} - \boldsymbol{r}',h)$ is a weighting function, namely the smoothing kernel, which smooths the values of $A(\boldsymbol{r}')$ in its influence area, defined by the so–called smoothing length h. Hence this smoothing length is also a measure for the introduced smoothness in the interpolated field $A_I(\boldsymbol{r})$. As a consequence, differentiability is achieved with the advantage that the derivative can be obtained analytically

$$\nabla A_I(\boldsymbol{r}) = \int\limits_{-\infty}^{\infty} A(\boldsymbol{r}')\nabla W(\boldsymbol{r} - \boldsymbol{r}',h)d^3r' , \tag{2.2}$$

see [Mon12], under the assumption of a complete support of the kernel.

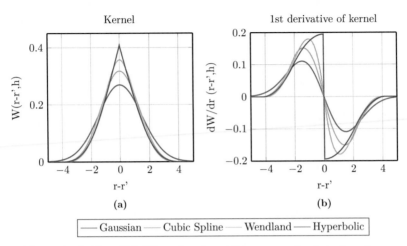

Figure (2.1): Plot of (a) different kernels and (b) corresponding derivatives, for a smoothing length of $h = 2.1$, normalized in 1D.

2.1.1 Kernel function

Different kernel types were used in the past, e.g. Gingold and Monaghan [Gin77] decided to use a Gaussian kernel, because the mean squared error of a desired density field was less, at least for a good enough resolution, compared to the spline kernels they investigated. In this sense the applied kernel reads [Mon92]

$$W_G(\boldsymbol{r} - \boldsymbol{r'}, h) = \frac{1}{h\sqrt{\pi}} e^{-q^2} \tag{2.3}$$

with

$$q = \frac{|\boldsymbol{r} - \boldsymbol{r'}|}{h} \tag{2.4}$$

and normalized in $1D$. However, the Gaussian kernel has no compact support, meaning especially when less particles are used for the resolution within its influence area, the approximation error becomes bigger. One has to keep in mind that the smoothing length cannot be freely chosen and increased, respectively, because the computational effort would

dramatically increase as well. A first investigation on kernel estimates for particle methods
was done by Gingold and Monaghan [Gin82].

The demands on the kernel are on the one hand

$$\int_{-\infty}^{\infty} W\left(\boldsymbol{r} - \boldsymbol{r}',h\right) d^3 r' = 1\,, \tag{2.5}$$

where usually a compact support is preferred to reduce the integration area and the
computational effort. On the other hand, in the limit of a vanishing smoothing length h,
the kernel shall become the Dirac delta function

$$\lim_{h \to 0} W\left(\boldsymbol{r} - \boldsymbol{r}',h\right) = \delta(\boldsymbol{r} - \boldsymbol{r}')\,. \tag{2.6}$$

A comparison of different kernel types is plotted in Fig. 2.1(a). A common kernel type is
the cubic spline [Mon05]

$$W_{CS}\left(\boldsymbol{r} - \boldsymbol{r}',h\right) = N \begin{cases} ((2-q)^3 - 4(1-q)^3) & \text{if } 0 \le q \le 1 \\ (2-q)^3 & \text{if } 1 \le q \le 2 \\ 0 & \text{else,} \end{cases} \tag{2.7}$$

with a normalization constant of $N = 1/(6h)$ in $1D$, $N = 15/(14\pi h^2)$ in $2D$ and $N = 1/(4\pi h^3)$ in $3D$. This kernel is widely used. But as recent publications show, it suffers
from a stronger clustering instability of the particles [Sze12a].

During the last years the quintic Wendland kernel [Wen95] has come more and more into
spotlight, because it leads to a more regularized particle distribution [Sze12a]

$$W_W\left(\boldsymbol{r} - \boldsymbol{r}',h\right) = N \begin{cases} \left(1 - \frac{q}{2}\right)^4 (2q+1) & \text{if } q < 2 \\ 0 & \text{else.} \end{cases} \tag{2.8}$$

Here the normalization constant is $N = 3/(4h)$ in $1D$, $N = 7/(4\pi h^2)$ in $2D$ and $N = 21/(16\pi h^3)$ in $3D$.

Another promising kernel function with respect to the clustering instability is the hyperbolic kernel as published by Yang et al. [Yan14]

$$
W_H\left(\boldsymbol{r}-\boldsymbol{r}',h\right) = N \begin{cases} q^3 - 6q + 6 & \text{if } 0 \le q < 1 \\ (2-q)^3 & \text{if } 1 \le q < 2 \\ 0 & \text{else,} \end{cases} \tag{2.9}
$$

with a normalization constant of $N = 1/(7h)$ for 1D, $N = 1/(3\pi h^2)$ for 2D and $N = 15/(62\pi h^3)$ for 3D.

The kernel derivative is then obtained through

$$
\nabla W\left(\boldsymbol{r}-\boldsymbol{r}',h\right) = \frac{\partial W}{\partial q}\frac{\partial q}{\partial r}\frac{\boldsymbol{r}-\boldsymbol{r}'}{|\boldsymbol{r}-\boldsymbol{r}'|} \tag{2.10}
$$

and illustrated in Fig. 2.1(b) for a one–dimensional test case.

2.1.2 Discrete interpolation

To build a numerical scheme the analytical interpolation has to be transformed. The discrete formulation is obtained by first expanding the integrand of Eq. (2.1) through a density quotient with $\rho/\rho = 1$, see [Ros09],

$$
A_I(\boldsymbol{r}) = \int_{-\infty}^{\infty} \frac{A(\boldsymbol{r}')}{\rho(\boldsymbol{r}')} W(\boldsymbol{r}-\boldsymbol{r}',h)\,\rho(\boldsymbol{r}')d^3r' \,. \tag{2.11}
$$

Then, to transform the continuous description to a discrete description, the integral in space is replaced by a sum over interpolation points accompanied by a change of the variable of integration. Hence, the term $\rho(\boldsymbol{r}')d^3r'$ is transformed into a new discrete mass element, representing the mass m_b of a hereby introduced particle b. The resulting description reads

$$
A(\boldsymbol{r}_a) = \sum_b \frac{m_b}{\rho_b} A_b W(\boldsymbol{r}_a - \boldsymbol{r}_b,h) \tag{2.12}
$$

with $A_b = A(\boldsymbol{r}_b)$ being used for convenience. A_b represents the field quantity A of particle b at the position $\boldsymbol{r}_b$.

2.1.3 Error estimation

To estimate the error introduced by this interpolation approach of Eq. (2.1), the physical field quantity $A(r')$ is expressed through a Taylor series. The interpolation approach $A_I(\boldsymbol{r})$ is a good approximation of $A(\boldsymbol{r})$ if the first, zeroth order term of the Taylor series remains, while higher order terms vanish and an acceptable error order is reached.

For simplicity here only a one–dimensional Taylor series expansion is made with respect to position r, which means r is constant. In this manner the general expansion reads

$$A(r') = A(r) + \frac{(r'-r)}{1!} \left.\frac{\partial A(r')}{dr'}\right|_{r'=r} + \frac{(r'-r)^2}{2!} \left.\frac{\partial^2 A(r')}{dr'^2}\right|_{r'=r} + \frac{(r'-r)^3}{3!} \left.\frac{\partial^3 A(r')}{dr'^3}\right|_{r'=r} + \dots \tag{2.13}$$

This series is now inserted in the interpolation of Eq. (2.1), according to [Mon82], and results in

$$A_I(r) = \int\limits_{-\infty}^{\infty} W(r-r',h) \left[A(r) - (r'-r) \left.\frac{\partial A(r')}{dr'}\right|_{r'=r} + \frac{(r-r')^2}{2!} \left.\frac{\partial^2 A(r')}{dr'^2}\right|_{r'=r} - \dots \right] dr'$$

$$= A(r) \underbrace{\int\limits_{-\infty}^{\infty} W(r-r',h)dr'}_{=1} + \left.\frac{\partial^2 A(r')}{dr'^2}\right|_{r'=r} \int\limits_{-\infty}^{\infty} \frac{(r-r')^2}{2!} W(r-r',h)dr' + \dots \tag{2.14}$$

Due to the fact that the kernel $W(r-r',h)$ is in general an even (symmetric) function, the terms with odd powers of $(r-r')$ vanish in the integration. Using Eq. (2.4) the kernel is formulated in a dimensionless way as introduced by [Qui06]

$$A_I(r) = A(r) + \underbrace{\left.\frac{\partial^2 A(r')}{dr'^2}\right|_{r'=r} \frac{h^2}{2!} \int\limits_{-\infty}^{\infty} q^2 W(q,h)h dq}_{\mathcal{O}(h^2)} + \dots \tag{2.15}$$

Hence, the error in the integration interpolant is of order $\mathcal{O}(h^2)$, which means this interpolation technique is at least as accurate as a linear interpolation. To increase accuracy one has to change the kernels in order to get rid of even higher order terms in Eq. (2.15) while preserving the zeroth moment. For instance, corrective formulations of the kernel exist to exactly describe polynomials of a certain degree [Bon99, Li96, Liu97].

Additionally to the above mentioned error, the SPH–interpolation also suffers from a "quadrature error", originating from the discrete description of Eq. (2.12), and an "anisotropy error", which is in general introduced due to irregular particle distributions. Therewith the total error reads [Tar15]

$$||e_{tot}|| \leq ||e_{int}|| + ||e_{quad}|| + ||e_{anis}||. \tag{2.16}$$

In this way the SPH error sum is composed of an analytical interpolation error e_{int} and the numerical errors e_{quad} and e_{anis}. The quadrature error scales with l_0/h, where l_0 is used for the initial particle spacing. Throughout this work, the smoothing length h is set to $h = 2.1 l_0$. This means the quotient l_0/h is usually constant and a representation of the amount of particles, which are used in one interpolation, see [Qui06]. With respect to the anisotropy error the work of [Tar15, Fat11] is referred to. They investigated the influence of different derivative operators for various inhomogeneous particle arrangements.

2.1.4 Derivatives

There are different ways how derivatives can be built in SPH. Here the methods are explained in detail, which are considered and applied in this thesis. Nevertheless, some recent modifications and improvements will be described afterwards at the end of this section.

First derivative

As previously mentioned, the advantage of this interpolation method is that the interpolation kernel is an analytically known function and therefore its derivative can also be analytically derived straight forward [Mon05, Ros09]. From Eq. (2.2) follows

$$\nabla A(\boldsymbol{r}_a) = \sum_b V_b A_b \nabla W(\boldsymbol{r}_{ab}, h), \tag{2.17}$$

where $\boldsymbol{r}_{ab} = \boldsymbol{r}_a - \boldsymbol{r}_b$ and $V_b = m_b/\rho_b$ is used for convenience. The disadvantage of this description is, it is sensitive towards particle disordering. For example, the derivative of a constant function $A(r) = c$ would in general not become zero, because the weighting is coming from the kernel derivatives and as long as the distances between particles are not the same, the weighting is not the same and therefore in case of an irregular particle arrangement the discrete derivative of a constant scalar field using Eq. (2.17) would not exactly become zero.

Another approach is to use a bump function Φ with $\nabla(\Phi A) = A\nabla\Phi + \Phi\nabla A$ and formulate the derivative this way

$$\nabla A \;=\; \frac{\nabla(\Phi A) - A\nabla\Phi}{\Phi}\,. \tag{2.18}$$

The discrete derivative therewith becomes

$$(\nabla A)_a \overset{(2.17)}{=} \frac{1}{\Phi_a}\left(\sum_b V_b\Phi_b A_b \nabla W_{ab} - A_a \sum_b \Phi_b V_b \nabla W_{ab}\right) \tag{2.19}$$

$$= \frac{1}{\Phi_a}\sum_b V_b\Phi_b\left(A_b - A_a\right)\nabla W_{ab}\,. \tag{2.20}$$

The other way is to use a bump function by $\nabla(A/\Phi) = (\Phi\nabla A - A\nabla\Phi)/\Phi^2$ in order to obtain

$$\nabla A \;=\; \Phi\nabla\left(\frac{A}{\Phi}\right) + \frac{A}{\Phi}\nabla\Phi \tag{2.21}$$

and correspondingly

$$(\nabla A)_a \overset{(2.17)}{=} \Phi_a \sum_b V_b \frac{A_b}{\Phi_b}\nabla W_{ab} + \frac{A_a}{\Phi_a}\sum_b V_b\Phi_b\nabla W_{ab} \tag{2.22}$$

$$= \sum_b V_b\left(\Phi_a\frac{A_b}{\Phi_b} + \Phi_b\frac{A_a}{\Phi_a}\right)\nabla W_{ab}\,. \tag{2.23}$$

If now a constant value is inserted for the bump function, say $\Phi = 1$, for the first case this results in

$$(\nabla A)_a = \sum_b V_b\left(A_b - A_a\right)\nabla W_{ab}\,, \qquad \text{the ``$-$''-formulation} \tag{2.24}$$

and for the second case in

$$(\nabla A)_a = \sum_b V_b\left(A_b + A_a\right)\nabla W_{ab}\,, \qquad \text{the ``$+$''-formulation.} \tag{2.25}$$

Both descriptions are valid and both have advantages and disadvantages, e.g. assuming an irregular particle arrangement, the "−"–formulation in contrast to the "+"–formulation indeed becomes zero for a constant field $A(r) = c$. The advantage of the latter formulation will be explained later, when conservation is considered.

Error estimation of the first derivative

To investigate the interpolation error with respect to the first derivative, the approach of Eq. (2.2) is verified by inserting the Taylor series of Eq. (2.13) and checked for occurring higher order terms [Qui06]. Because the approach yields a description for the first derivative, the zeroth order term of the Taylor series is expected to vanish, whereas the second term shall remain and the occurring higher order terms define the interpolation error. For simplicity, the expansion is again reduced to a one–dimensional consideration

$$\int_{-\infty}^{\infty} A(r') \frac{\partial W}{\partial r}(r - r',h)\, dr' =$$

$$- \int_{-\infty}^{\infty} \left[A(r) + (r' - r) \frac{\partial A(r')}{\partial r'}\bigg|_{r'=r} + \frac{(r' - r)^2}{2!} \frac{\partial^2 A(r')}{\partial r'^2}\bigg|_{r'=r} \right.$$

$$\left. + \frac{(r' - r)^3}{3!} \frac{\partial^3 A(r')}{\partial r'^3}\bigg|_{r'=r} + \dots \right] \frac{\partial W}{\partial r'}(r - r',h)\, dr' . \quad (2.26)$$

Placing the constant derivatives outside the brackets leads to

$$\frac{\partial A(r')}{\partial r'}\bigg|_{r'=r} \int_{-\infty}^{\infty} (r - r') \frac{\partial W}{\partial r'}(r - r',h)\, dr'$$

$$- \frac{1}{2!} \frac{\partial^2 A(r')}{\partial r'^2}\bigg|_{r'=r} \int_{-\infty}^{\infty} (r - r')^2 \frac{\partial W}{\partial r'}(r - r',h)\, dr'$$

$$+ \frac{1}{3!} \frac{\partial^3 A(r')}{\partial r'^3}\bigg|_{r'=r} \int_{-\infty}^{\infty} (r - r')^3 \frac{\partial W}{\partial r'}(r - r',h)\, dr' - \dots . \quad (2.27)$$

The first term in brackets, namely $A(r)$, in Eq. (2.26) vanished, because the integral with the odd kernel derivative becomes zero. Integration by parts gives rise to

$$\frac{\partial A(r')}{dr'}\bigg|_{r'=r} \underbrace{\int\limits_{-\infty}^{\infty} W(r-r',h)\,dr'}_{=1}$$

$$-\frac{\partial^2 A(r')}{dr'^2}\bigg|_{r'=r} \int\limits_{-\infty}^{\infty} (r-r')W(r-r',h)\,dr'$$

$$+\frac{1}{2}\frac{\partial^3 A(r')}{dr'^3}\bigg|_{r'=r} \int\limits_{-\infty}^{\infty} (r-r')^2 W(r-r',h)\,dr' - \dots . \quad (2.28)$$

The first line of Eq. (2.28) represents the desired derivative, and the second line vanishes as every other term with an odd power of $(r - r')$ does. Therefore, the interpolation error of the first derivative reduces to

$$\frac{\partial}{\partial r}A_I(r) = \frac{\partial A(r')}{dr'}\bigg|_{r'=r} + \underbrace{\frac{h^2}{2}\frac{\partial^3 A(r')}{dr'^3}\bigg|_{r'=r} \int\limits_{-\infty}^{\infty} q^2 W(q,h)h\,dq}_{\mathcal{O}(h^2)} + \dots . \quad (2.29)$$

This result is not surprising as the kernel estimate already had an error in second order $\mathcal{O}(h^2)$ and the first derivative is an analytical derivative of the kernel estimate. Therefore, no additional interpolation error should be introduced. Nevertheless, it reveals the general procedure to obtain the order of the integral approximation due to the applied approach with the corresponding interpolation technique. Note that in a discrete formulation also a quadrature error and an anisotropy error is introduced as mentioned previously.

Second derivative

The straight forward formulation of the second derivative

$$(\Delta A)_a = \sum_b V_b A_b \Delta W(\boldsymbol{r}_{ab},h) \quad (2.30)$$

is strongly depending on the particle position and the applied kernel. It may for instance become negative, only depending on the particle distribution, cf. Fig. 2.2. All utilized

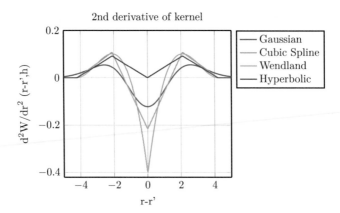

Figure (2.2): Second derivative of interpolation kernel, cf. Fig. 2.1.

kernels except the hyperbolic one of Eq. (2.9) have a partially negative second derivative. Therefore, in an exemplary case of a heat conduction problem, this formulations may give rise to an unphysical heat flux from a colder to a warmer particle, just because of the arrangement of two particles and not because of the temperature profile [Mon05]. To omit the usage of the second kernel derivative, the first derivative can be applied two times in a row

$$(\Delta A)_a = \sum_b V_b \left(\sum_c V_c A_c \nabla W(\boldsymbol{r}_{bc}, h) \right) \cdot \nabla W(\boldsymbol{r}_{ab}, h). \tag{2.31}$$

The disadvantage of this approach is that the smoothing is applied two times, meaning the influence area is doubled and thereby the boundary and ghost particles area need to be twice as thick as regular [Bas09], which complicates the formulation of accurate boundary conditions. Therefore, a different formulation of the second derivative is used, according to [Bro85], which can be considered as an outer SPH–derivative, cf. Eq. (2.25), and an inner finite difference approximation

$$(\nabla (\Phi \nabla A))_a = \sum_b V_b \frac{(\Phi_a + \Phi_b)(A_a - A_b)}{|\boldsymbol{r}_{ab}|^2 + \zeta^2} \boldsymbol{r}_{ab} \cdot \nabla W_{ab}. \tag{2.32}$$

Here ζ is a small numerical constant, preventing the denominator to become zero and set to $\zeta = 0.01h$. This formulation also fulfills the condition, that the second derivative of a constant field $A(\boldsymbol{r}) = c$ vanishes and that the truncation error scales with $\mathcal{O}(h^2)$, which is proved in the following.

Error estimation of the second derivative

In analogy to the previous investigations the integral approximation is analyzed using an approach for the second derivative. The analytical approach, which leads to the discretized version of Eq. (2.32), is given in [Bro85]. For simplicity $\varPhi = 1$ is assumed

$$\frac{\partial^2}{\partial r^2} \tilde{A}_I(r) = 2 \int\limits_{-\infty}^{\infty} \frac{[A(r) - A(r')]}{(r - r')} \frac{\partial W}{\partial r}(r - r',h)\,dr'\,. \tag{2.33}$$

Inserting the Taylor series of Eq. (2.13) gives rise to

$$- 2 \int\limits_{-\infty}^{\infty} \frac{1}{(r - r')} A(r) \frac{\partial W}{\partial r'}(r - r',h)\,dr'$$

$$+ 2 \int\limits_{-\infty}^{\infty} \frac{1}{(r - r')} \left[A(r) - (r - r') \left.\frac{\partial A(r')}{dr'}\right|_{r'=r} + \frac{(r' - r)^2}{2!} \left.\frac{\partial^2 A(r')}{dr'^2}\right|_{r'=r} \right.$$

$$\left. - \frac{(r - r')^3}{3!} \left.\frac{\partial^3 A(r')}{dr'^3}\right|_{r'=r} + \frac{(r - r')^4}{4!} \left.\frac{\partial^4 A(r')}{dr'^4}\right|_{r'=r} - \dots \right] \frac{\partial W}{\partial r'}(r - r',h)\,dr'\,. \tag{2.34}$$

The first terms with $A(r)$ cancel each other and the second term with $\partial A(r')/\partial r'$ vanishes because of the integration with the odd kernel derivative, as every other derivative term with an odd power does, hence

$$\left.\frac{\partial^2 A(r')}{dr'^2}\right|_{r'=r} \int\limits_{-\infty}^{\infty} (r - r') \frac{\partial W}{\partial r'}(r - r',h)\,dr'$$

$$+ \frac{2}{4!} \left.\frac{\partial^4 A(r')}{dr'^4}\right|_{r'=r} \int\limits_{-\infty}^{\infty} (r - r')^3 \frac{\partial W}{\partial r'}(r - r',h)\,dr' + \dots \tag{2.35}$$

is obtained. Integrating by parts and writing the kernel in a dimensionless way gives rise to

$$
\frac{\partial^2}{\partial r^2} \tilde{A}_I(r) = \underbrace{\left.\frac{\partial^2 A(r')}{dr'^2}\right|_{r'=r} \int\limits_{-\infty}^{\infty} W(q,h)hdq}_{=1} + \underbrace{\frac{h^2}{4} \left.\frac{\partial^4 A(r')}{dr'^4}\right|_{r'=r} \int\limits_{-\infty}^{\infty} q^2 W(q,h)hdq + \dots}_{=\mathcal{O}(h^2)},
$$

$$(2.36)$$

which illustrates, that the interpolation error in this formulation of the second derivative is again of order $\mathcal{O}(h^2)$. Again in a discrete formulation a quadrature error and an anisotropy error complete the error consideration.

2.1.5 Consistency

To explain consistency the description of Basa et al. [Bas09] is referred to:

> "The order of consistency of an operation represents the order of the polynomial function on which the operation evaluates exactly, when using values at discrete particle positions."

These discrete particle positions in general represent arbitrary particle distributions [Qui06]. This is why the standard SPH approach, as described above, is not even zeroth order consistent. Under general conditions the kernel sum may be $\sum_b V_b W(r_{ab},h) \neq 1$, e.g. due to the anisotropy error. Moreover, the same problem arises because of the particle deficiency at a boundary.

2.1.6 Corrected SPH

Shepard correction

To compensate for a particle deficiency and an irregular particle arrangement in the discretized description the so–called Shepard correction [She68] is applied. This modification can be considered as a renormalization of the kernel sum by the actually present particles to ensure $\sum_b V_b \widetilde{W}(r_{ab},h) = 1$ through an adjusted kernel formulation

$$
\widetilde{W}(r_{ab},h) = \frac{W(r_{ab},h)}{\varepsilon_a},
$$

$$(2.37)$$

with normalization $\varepsilon_a = \sum_b V_b W(r_{ab},h)$. The derivative of the kernel therewith becomes

$$\nabla \widetilde{W}(r_{ab},h) = \frac{\nabla W_{ab} - W_{ab} \sum_b V_b \nabla W_{ab}/\varepsilon_a}{\varepsilon_a} . \tag{2.38}$$

In the last step, the quotient rule was applied and $W_{ab} = W(r_{ab},h)$ is used for clarity. This modification of the kernel guarantees zeroth–order consistency, cf. [Bas09, Bon99]. But rather than describing the antiderivative, a good approximation of the differential operator is necessary when it comes to partial differential equations.

Mixed kernel and gradient correction

This is why an additional correction of the gradient operator is used and thereby a combination of Shepard correction and gradient correction is utilized, following the work of Bonet and Lok [Bon99]. The idea is to introduce a correction on the derivative operator

$$\widetilde{\nabla}\widetilde{W}(r_{ab},h) = L_a \nabla \widetilde{W}_{ab} \tag{2.39}$$

to exactly evaluate the gradient of any linear vector field. Therefore the condition

$$\sum_b V_b r_{ba} \otimes \widetilde{\nabla}\widetilde{W}_{ab} = I \tag{2.40}$$

has to be fulfilled, where the "minus"–formulation of Eq. (2.24) is applied (the simplified corresponding equation in 1D would be $\partial_x x = 1$). Note that due to the previous zeroth–order correction, as r_a is constant, it would also be possible to change $r_{ba} \to r_b$. With Eq. (2.39) and Eq. (2.40) the correction matrix becomes

$$L_a = \left(\sum_b V_b \nabla \widetilde{W}_{ab} \otimes r_{ba} \right)^{-1} . \tag{2.41}$$

Therewith, the first order correction for arbitrary particle distributions is accomplished. Note that this formulation also ensures rotational invariance as the gradient is calculated and the result is obtained independent on the particle arrangement. For a deeper insight into this method and validations see the PhD thesis of F. Keller [Kel14].

Other corrections and higher order methods

Several higher order methods exist to increase consistency and accuracy of SPH calculations. Derivative approximations can be corrected for example by considering Taylor series expansion with higher–order terms, see Chen et al. [Che99] and Zhang and Batra [Zha04], respectively. The latter uses a Taylor series expansion considering terms up to the order of $(m + 1)$, to obtain a kernel estimate of mth order and a first and second derivative with a consistency of order $(m - 1)$ and $(m - 2)$. Although this approach is interesting, because any order can be realized, the computational effort dramatically increases, as already for $m = 2$ a linear equation system of size $10x10$ for each particle has to be solved and the error of the integral approximation for the derivatives is even worse.

At this point it is worth mentioning other approaches for higher order systems. There are several publications based on a Moving Least–Squares (MLS) approach, e.g. see Liu et al. [Liu97], Li and Liu [Li96] and Bilotta et al. [Bil11]. The latter illustrates the advantage of this approach. It is not necessary to build derivatives of the kernel function to obtain the derivative of a physical quantity. Therefore conventional SPH rather suffers from tensile instability than the MLS approach [Bil11]. With tensile instability and pairing instability respectively [Fan06, Jia10] the unphysical behavior of particle clustering and as a consequence a potential breakup of the whole system is addressed. This is also why hyperbolic kernel functions recently came into spotlight [Yan14].

Yet another method to stabilize the discrete system is to introduce Riemann solvers [Inu02, GG12, Raf12, Sir13].

2.2 Conservation laws

From Noether's theorem it is known that continuous symmetries in physical systems have corresponding conservation laws. This is why translation invariance gives rise to conservation of linear momentum, rotation invariance gives rise to conservation of angular momentum and conservation of energy is achieved by translation invariance in time.

2.2.1 The Euler equations

For illustration the Euler equations are introduced, which are the most simple set of equations in fluid dynamics. They describe an adiabatic single–phase flow for ideal, inviscid fluids. The Lagrangian description is given by

$$\begin{cases} \frac{d\rho}{dt} = \rho \nabla \cdot \boldsymbol{v} \\ \frac{d\boldsymbol{v}}{dt} = -\frac{\nabla p}{\rho} + \boldsymbol{g} \\ \frac{du}{dt} = -\frac{p}{\rho} \nabla \cdot \boldsymbol{v} \,. \end{cases} \qquad (2.42)$$

Here, the vectors $\boldsymbol{v}$ and $\boldsymbol{g}$ denote the velocity and the gravitational acceleration. The scalars p and u are the pressure and the internal energy of the system. The set of equations consists of a mass balance, namely the continuity equation, a momentum balance, built by a pressure gradient plus gravitation and an energy balance. It it obvious to see, that in case of an incompressible flow, the continuity equation yields $\nabla \cdot \boldsymbol{v} = 0$. Consequently the energy balance becomes $du/dt = 0$ and from an analytical and physical point of view the remaining challenge is to solve the momentum balance.

From a numerical point of view the conservation of mass, energy and momentum is not in general fulfilled, as this depends on the discretization of the differential operators.

The following conservation considerations are made for complete, closed physical systems.

2.2.2 Linear momentum

In terms of the linear momentum $\boldsymbol{P} = m\boldsymbol{v}$ this means the equation

$$\frac{\partial \boldsymbol{P}}{\partial t} = m \frac{\partial \boldsymbol{v}}{\partial t} \overset{!}{=} 0 \qquad (2.43)$$

has to be fulfilled under the assumption of constant masses. In the SPH formulation the sum over all particles in the complete system has to be considered [Ros09, Bon99]

$$\sum_a \underbrace{m_a \frac{d\boldsymbol{v}_a}{dt}}_{\boldsymbol{F}_a = \sum_b \boldsymbol{F}_{ab}} = \sum_a \sum_b \boldsymbol{F}_{ab} = \frac{1}{2} \sum_a \sum_b (\boldsymbol{F}_{ab} + \boldsymbol{F}_{ba}) = 0 \,. \qquad (2.44)$$

Here $\boldsymbol{F}_{ab}$ denotes the force of particle b on particle a and vice versa. Hence, the validity of this equation is confirmed if $\boldsymbol{F}_{ab} = -\boldsymbol{F}_{ba}$.

Let's have a look at the momentum balance of the Euler equations (2.42). As we here consider the conservation of linear momentum we can get rid of the term "$+\boldsymbol{g}$", because gravitation represents a homogeneous force field, which anyway fulfills the condition of translation invariance and therefore conservation of linear momentum is already given for this contribution. Using the "−"–formulation of Eq. (2.24) gives rise to

$$\boldsymbol{F}_{ab}^- = -V_a V_b \left(p_b - p_a \right) \nabla W_{ab} , \tag{2.45}$$

whereas the "+"–formulation of Eq. (2.25) leads to

$$\boldsymbol{F}_{ab}^+ = -V_a V_b \left(p_b + p_a \right) \nabla W_{ab} . \tag{2.46}$$

Due to the odd kernel derivative $\nabla W_{ab} = -\nabla W_{ba}$, cf. Fig. 2.1b, an interchange of a and b reveals

$$\boldsymbol{F}_{ab}^- = \boldsymbol{F}_{ba}^- \text{ and} \tag{2.47}$$
$$\boldsymbol{F}_{ab}^+ = -\boldsymbol{F}_{ba}^+ . \tag{2.48}$$

Therefore the "+"–formulation is obviously conserving linear momentum. Note that this is just a proof, whether a formulation is conserving a quantity and not a counterproof, as other argumentations may exist, which confirm Eq. (2.44) as well.

2.2.3 Angular momentum

With regard to angular momentum $\boldsymbol{L} = \boldsymbol{r} \times \boldsymbol{P}$ the conservation thereof is given if

$$\frac{\partial \boldsymbol{L}}{\partial t} = \boldsymbol{M} = \boldsymbol{r} \times \boldsymbol{F} \stackrel{!}{=} \boldsymbol{0} \tag{2.49}$$

is fulfilled. This means in the SPH formulation by taking the sum over all particles in the closed system conservation is obtained through

$$\sum_a \boldsymbol{r}_a \times \boldsymbol{F}_a = \sum_a \sum_b \boldsymbol{r}_a \times \boldsymbol{F}_{ab} = \frac{1}{2} \sum_a \sum_b \left(\boldsymbol{r}_a \times \boldsymbol{F}_{ab} + \boldsymbol{r}_b \times \boldsymbol{F}_{ba} \right) \tag{2.50}$$

$$= \frac{1}{2} \sum_a \sum_b \left(\boldsymbol{r}_a \times \boldsymbol{F}_{ab} - \boldsymbol{r}_b \times \boldsymbol{F}_{ab} \right) = \frac{1}{2} \sum_a \sum_b \left(\boldsymbol{r}_{ab} \times \boldsymbol{F}_{ab} \right) = \boldsymbol{0} , \tag{2.51}$$

e.g. if $\boldsymbol{r}_{ab}$ and $\boldsymbol{F}_{ab}$ are collinear. This is achieved for instance by building the force with the direction of the kernel derivative, as realized in Eq. (2.45) and Eq. (2.46). Due to the fact that

$$\nabla W_{ab} = \frac{\partial W(q,h)}{\partial q} \frac{\partial q(r_{ab},h)}{\partial r_{ab}} \hat{r}_{ab} \,, \tag{2.52}$$

with $\hat{r}_{ab}$ as the unit vector, the cross product of both vectors r_{ab} and F_{ab} in this case is the zero vector, which confirms angular momentum conservation.

2.2.4 Mass

Usually mass is conserved in an analytical physical framework by applying the continuity equation, which is in Lagrangian formulation given by

$$\frac{d\rho}{dt} + \rho \nabla \cdot v = 0 \,. \tag{2.53}$$

Here the right hand side, which is responsible for the generation of mass, is explicitly set to zero. In a numerical approach, the semi–discrete equation for time integration then becomes

$$\frac{d\rho_a}{dt} = -\rho_a \sum_b V_b \left(v_b \pm v_a \right) \cdot \nabla W_{ab} \,, \tag{2.54}$$

which does in a general sense not conserve mass, as the velocity field can be of any order and therefore even with a corrected approach for a certain order of consistency, the order of the velocity field might be even higher. As a consequence, exact reproduction is not guaranteed and mass is not conserved in a numerical sense. To overcome this issue, the density field is not estimated through a differential equation but with a summation over the present particles in terms of Eq. (2.12). With $A_b = \rho_b$ this leads to

$$\rho_a = \sum_b m_b W_{ab} \,. \tag{2.55}$$

Note that this formulation may also introduce an interpolation, quadrature and anisotropy error, but with respect to Eq. (2.54), this error does not sum up over time. Moreover, if the masses m_b are constant within one phase, which is usually the case for incompressible flows, the previously mentioned error terms vanish by using the corrected approach of Eq. (2.37). In this way the kernel has to be replaced by $W_{ab} \to \widetilde{W}_{ab}$. Additionally, in a multi–phase environment the formulation of the density approximation is adjusted to get rid of the smooth density approximation at a two–phase interface and ensure again a density jump for a better interface sharpness according to [Hu06]. This gives rise to

$$\rho_a = m_a \sum_b \widetilde{W}_{ab} . \tag{2.56}$$

It is important to note that this formulation decouples the density approximation from the local particle arrangement, because the aim of the corrected approach is to get rid of the anisotropy error. But this is not in every case the preferred or desired evaluation, as will be explained in the section about the pressure Poisson equation.

2.2.5 Energy

To obtain the energy balance we start with the first law of thermodynamics

$$dU = dQ - pdV . \tag{2.57}$$

Here, U, Q and V are internal energy, adiabatic energy (heat) and volume. In case of the adiabatic Euler equations dQ vanishes. According to Rosswog [Ros09] the thermodynamic quantities are written on a "per mass basis". Therefore, U becomes u and $dV \rightarrow d(1/\rho) = -1/\rho^2 d\rho$. The balance of the internal energy then reads

$$\frac{du}{dt} = \frac{p}{\rho^2} \frac{d\rho}{dt} , \tag{2.58}$$

which leads to

$$\frac{du}{dt} = -\frac{p}{\rho} \nabla \cdot \boldsymbol{v} . \tag{2.59}$$

Using the continuity equation and by applying the SPH discretization

$$\frac{du_a}{dt} = -\frac{p_a}{\rho_a} \sum_b V_b \left(\boldsymbol{v}_b \pm \boldsymbol{v}_a \right) \cdot \nabla W_{ab} \tag{2.60}$$

is obtained.

The total energy balance is then given through

$$\frac{dE}{dt} = \frac{d}{dt} \sum_a \Big(\underbrace{m_a u_a}_{\substack{\text{internal} \\ \text{energy of } a}} + \underbrace{\frac{1}{2} m_a v_a^2}_{\substack{\text{kinetic} \\ \text{energy of } a}} \Big) = \sum_a m_a \left(\frac{du_a}{dt} + \boldsymbol{v}_a \cdot \frac{d\boldsymbol{v}_a}{dt} \right) . \tag{2.61}$$

Using Eq. (2.60) and a momentum balance of $d\boldsymbol{v}/dt = -1/\rho\, \nabla p$, the SPH formulation of the total energy becomes

$$\frac{dE}{dt} = \sum_a m_a \left(-\frac{p_a}{\rho_a} \sum_b V_b \left(\boldsymbol{v}_b \pm \boldsymbol{v}_a \right) \cdot \nabla W_{ab} - \boldsymbol{v}_a \cdot \frac{1}{\rho_a} \sum_b V_b \left(p_b \pm p_a \right) \nabla W_{ab} \right)$$

$$= \sum_a \left(-p_a \sum_b V_a V_b \left(\boldsymbol{v}_b \pm \boldsymbol{v}_a \right) \cdot \nabla W_{ab} - \boldsymbol{v}_a \cdot \sum_b V_a V_b \left(p_b \pm p_a \right) \nabla W_{ab} \right) \tag{2.62}$$

$$= - \sum_a \sum_b V_a V_b \left(p_a \boldsymbol{v}_b + p_b \boldsymbol{v}_a \right) \cdot \nabla W_{ab} = 0 .$$

In the last step a smart choice of the "-/+"–formulation yields an energy conserving description, because again due to the odd kernel derivative, the energy contributions cancel pairwise.

2.3 Time integration of the incompressible scheme

To solve the Navier–Stokes equations for incompressible, Newtonian fluids the decomposition of Chorin [Cho68] is used. The basic idea of the incompressible SPH approach (ISPH) is to split the momentum balance for the time integration into two parts

$$\frac{D\boldsymbol{v}}{Dt} = \underbrace{\frac{\mu}{\rho} \Delta \boldsymbol{v} + \boldsymbol{g}}_{\boldsymbol{a}_{\mathrm{I}}} \underbrace{- \frac{1}{\rho} \nabla p}_{\boldsymbol{a}_{\mathrm{II}}} . \tag{2.63}$$

The part $\boldsymbol{a}_{\mathrm{I}}$ is accomplished in the predictor step whereas the part $\boldsymbol{a}_{\mathrm{II}}$ is defined through the previous step and the incompressibility condition

$$\frac{D\rho}{Dt} = \rho \nabla \cdot \boldsymbol{v} = 0 . \tag{2.64}$$

The last step is therefore called corrector step and is in particular described in combination with the time integration scheme.

First a common Euler scheme is elucidated and then in the context of ISPH a new Leap–Frog scheme is introduced, which is preferable over the previous one, because of the truncation error and its conservation properties.

2.3.1 Euler scheme

Cummins and Rudman [Cum99] first published the incompressible predictor corrector scheme for SPH by using a first–order Euler scheme. The general explicit Euler scheme is hereby given through

$$v_{n+1} = v_n + \Delta t\, a_n \quad \text{and} \tag{2.65}$$

$$r_{n+1} = r_n + \Delta t\, v_n, \tag{2.66}$$

where n denotes the discrete time step, Δt the time step interval and $a = a_\mathrm{I} + a_\mathrm{II}$ the complete acceleration according to the momentum balance of Eq. (2.63).

In order to obtain an incompressible flow field for each upcoming time step, Eq. (2.64) has to be fulfilled. By this the divergence of Eq. (2.65) has to equal zero. Following the derivation of Shao and Lo [Sha03] this means

$$\nabla \cdot v_{n+1} = \nabla \cdot (v_n + \Delta t\, a_\mathrm{I} + \Delta t\, a_\mathrm{II}) \overset{!}{=} 0. \tag{2.67}$$

Here the decomposition comes into play. In the first step an intermediate velocity is built by

$$v_{int} = v_n + \Delta t\, a_\mathrm{I}. \tag{2.68}$$

Then by inserting v_{int} into Eq. (2.67), the pressure Poisson equation is obtained

$$\frac{1}{\Delta t} \nabla \cdot v_{int} = \nabla \cdot \left(\frac{1}{\rho} \nabla P \right). \tag{2.69}$$

This implicit equation in space is solved to get the quantity which is attributed to an incompressible flow behavior. Through this projection a reasonable pressure field is obtained, which is in turn used to accomplish the second step

$$v_{n+1} = v_{int} + \Delta t\, a_{\mathrm{II}} = v_{int} - \Delta t \left(\frac{1}{\rho} \nabla P \right) . \tag{2.70}$$

Then by [Cum99, Sha03] the positions are updated through

$$r_{n+1} = r_n + \Delta t \left(\frac{v_n + v_{n+1}}{2} \right) . \tag{2.71}$$

2.3.2 Leap–Frog scheme

As a new time integration scheme in ISPH the second order Leap–Frog method is introduced
with

$$v_{n+0.5} = v_{n-0.5} + \Delta t\, a_n \quad \text{and} \tag{2.72}$$
$$r_{n+1} = r_n + \Delta t\, v_{n+0.5} . \tag{2.73}$$

This scheme is not self–starting, therefore the first step is made by

$$v_{0.5} = v_0 + 0.5\, \Delta t\, a_0. \tag{2.74}$$

The incompressible SPH scheme is built around the Leap–Frog scheme. Note that the
Leap–Frog scheme for weakly compressible SPH is also used in Liu and Liu [Liu03] and
Pan et al. [Pan13] but was introduced to ISPH by Huber et al. [Hub16b].
To obtain the predictor–corrector scheme for the Leap–Frog algorithm the same decomposition of Eq. (2.63) is used. The corresponding equations are then given by

$$v_{int} = v_{n-0.5} + \Delta t\, a_{\mathrm{I}} , \tag{2.75}$$
$$\frac{1}{\Delta t} \nabla \cdot v_{int} = \nabla \cdot \left(\frac{1}{\rho} \nabla P \right) , \tag{2.76}$$
$$v_{n+0.5} = v_{int} + \Delta t\, a_{\mathrm{II}} \quad \text{and} \tag{2.77}$$
$$r_{n+1} = r_n + \Delta t\, v_{n+0.5} . \tag{2.78}$$

2.3.3 Time step criteria

In order to obtain a converging solution in time, for stability reasons several criteria must
be fulfilled regarding the maximal time step size. On the one hand the Courant–Friedrichs–

Lewy (CFL) criterion

$$\Delta t \leq 0.1 \frac{l_0}{|\boldsymbol{v}|_{max}} \tag{2.79}$$

with $|\boldsymbol{v}|_{max}$ as the maximum velocity in the computational domain must be adhered. On the other hand, for viscous systems, an additional viscous diffusion condition

$$\Delta t \leq 0.1 \frac{l_0^2}{\eta_{max}} \tag{2.80}$$

with $\eta = \mu/\rho$ as kinematic viscosity and with η_{max}, denoting the maximum kinematic viscosity of all involved particles, has to be adhered.

The CFL criterion is applied twice, once in the predictor step with $|\boldsymbol{v}|_{max} = max\left(|\boldsymbol{v}_{int}|\right)$ to take care of external forces and surface tension forces respectively and once after the corrector step for the complete time step interval.

2.4 Common forms of derivatives in the momentum balance

In literature many different formulations for the gradient and Laplacian operators are used. Through the approach of the bump function as shown in Eqs. (2.18) and (2.21) along with the chosen value for Φ different formulations for derivatives can be derived. Some common descriptions are now highlighted.

Monaghan [Mon85, Mon92]

Through the historic evolution a very common approach for the discrete formulation of a gradient field is given in Eq. (2.81) and a rather empirical approach for the viscous formulation is given in Eq. (2.82).
Gradient

$$\left(\frac{1}{\rho}\nabla p\right)_a = \sum_b m_b \left(\frac{p_a}{\rho_a^2} + \frac{p_b}{\rho_b^2}\right) \nabla_a W_{ab} \tag{2.81}$$

Laplacian

$$(\nabla(\eta\nabla\cdot\boldsymbol{v}))_a \approx \sum_b m_b \Pi_{ab} \nabla_a W_{ab} \tag{2.82}$$

$$\Pi_{ab} = \begin{cases} \frac{\alpha c \tilde{\mu}_{ab} + \beta \tilde{\mu}_{ab}^2}{\bar{\rho}_{ab}}, & \text{if } \boldsymbol{v}_{ab}\cdot\boldsymbol{r}_{ab} < 0; \\ 0, & \text{otherwise} \end{cases} \tag{2.83}$$

$$\tilde{\mu}_{ab} = \frac{h\boldsymbol{v}_{ab}\cdot\boldsymbol{r}_{ab}}{|\boldsymbol{r}_{ab}|^2 + \zeta^2} \tag{2.84}$$

Here α and β are weighting coefficients and can be adjusted to the needs of the calculation, e.g. in [Mon85] α is constant and set to $\alpha = 1$, whereas $\beta = 0$. c is the speed of sound. Despite the other formulations below, here the idea behind this viscous model was not to discretize the viscous formulation of the Navier–Stokes equations but first to introduce an artificial viscosity to model shock phenomena and make the system more stable. Accordingly, a semi–empirical model for physical viscosity was in this way introduced [Mon12].

An advantage of these formulations is, they are both conserving linear and angular momentum.

Morris et al. [Mor97]

The discretization of the pressure gradient was adopted from the first publications of Monaghan. This formulation proved to be reliable for single–phase simulations or multi–phase simulations with low density ratios of almost ≈ 1 [Mor97, Cum99]. The numerical formulations in this work are given by:

Gradient

$$\left(\frac{1}{\rho}\nabla p\right)_a = \sum_b m_b \left(\frac{p_a}{\rho_a^2} + \frac{p_b}{\rho_b^2}\right)\nabla_a W_{ab} \tag{2.85}$$

Laplacian

$$(\nabla(\eta\nabla\cdot\boldsymbol{v}))_a = \sum_b \frac{m_b(\mu_a + \mu_b)\boldsymbol{r}_{ab}\cdot\nabla_a W_{ab}}{\rho_a\rho_b(|\boldsymbol{r}_{ab}|^2 + \zeta^2)}\boldsymbol{v}_{ab} \tag{2.86}$$

Eq. (2.86) represents a more realistic form of physical viscosity and the discretization is based on the derivation of a second derivative for heat conduction [Bro85]. It produces better results compared to the Monaghan formulation but with the disadvantage of conserving only approximately angular momentum [Mor97].

Shao and Lo [Sha03]

In this paper a single–phase system with a free surface is considered. For this purpose the discrete operators are given by:
Gradient

$$\left(\frac{1}{\rho}\nabla p\right)_a = \sum_b m_b \left(\frac{p_a}{\rho_a^2} + \frac{p_b}{\rho_b^2}\right) \nabla_a W_{ab} \tag{2.87}$$

Laplacian

$$(\nabla(\eta\nabla\cdot\boldsymbol{v}))_a = \sum_b \frac{4m_b(\mu_a + \mu_b)\boldsymbol{r}_{ab}\cdot\nabla_a W_{ab}}{(\rho_a + \rho_b)^2(|\boldsymbol{r}_{ab}|^2 + \zeta^2)}\boldsymbol{v}_{ab} \tag{2.88}$$

The viscous formulation is derived from the stress tensor but has also the drawback of an approximate conservation of angular momentum.

Adami et al. [Ada10]

The density weighted formulation of the pressure gradient was already published in [Hu07]. The advantage of this approach is, it allows to handle simulations with higher density ratios compared to the previous formulations through the modified slope at a two–phase interface. The numerical formulations are given by:
Gradient

$$\left(\frac{1}{\rho}\nabla p\right)_a = \frac{1}{m_a}\sum_b \left(V_a^2 + V_b^2\right)\frac{\rho_b p_a + \rho_a p_b}{\rho_a + \rho_b}\nabla_a W_{ab} \tag{2.89}$$

Laplacian

$$(\nabla(\eta\nabla\cdot\boldsymbol{v}))_a = \frac{1}{m_a}\sum_b \left(V_a^2 + V_b^2\right)\frac{2\mu_a\mu_b}{\mu_a + \mu_b}\frac{\boldsymbol{r}_{ab}\cdot\nabla_a W_{ab}}{|\boldsymbol{r}_{ab}|^2 + \zeta^2}\boldsymbol{v}_{ab} \tag{2.90}$$

This viscous discretization was also published in [Hu06]. The disadvantage here is again that the viscous formulation is only approximately conserving angular momentum. Nevertheless, this whole approach proves to be very promising especially with respect to high density ratios. A further advantage of this approach will be highlighted in the context of the boundary conditions and the viscous flow in a channel.

Szewc et al.[Sze12a]

The discrete formulation of the pressure gradient here was already published by Morris [Mor00] and Colagrossi and Landrini [Col03], but the Laplacian formulation represents a new discretization:

Gradient

$$\left(\frac{1}{\rho}\nabla p\right)_a = \sum_b m_b \left(\frac{p_a + p_b}{\rho_a \rho_b}\right) \nabla_a W_{ab} \tag{2.91}$$

Laplacian

$$\left(\nabla(\eta\nabla \cdot \boldsymbol{v})\right)_a = \sum_b 8m_b \left(\frac{\eta_a + \eta_b}{\rho_a + \rho_b} \frac{\boldsymbol{v}_{ab} \cdot \boldsymbol{r}_{ab}}{|\boldsymbol{r}_{ab}|^2 + \zeta^2}\right) \nabla_a W_{ab} \tag{2.92}$$

The advantage of this Laplacian formulation is, it is conserving both linear and angular momentum and therefore first became the model of choice in this work. The drawbacks will also be revealed in the context of the boundary conditions and the viscous flow in a channel.

3 Two–phase model

In order to describe laminar two–phase flow through a capillary, as illustrated in Fig. 3.1, several phenomena must be included in the physical model. Usually a momentum balance, a mass balance and an energy balance are necessary as a set of coupled equations. This set of equations is posed for each phase and the equations differ by the fluid properties like density, viscosity, etc. .

A general conservation law can be derived by balancing a quantity $\rho_k \psi_k$ in the material control volume V with surface A through [Ish11]

$$\frac{D}{Dt} \int_V \rho_k \psi_k \, dV = - \oint_A \boldsymbol{n}_k \cdot \boldsymbol{J}_k \, dA + \int_V \rho_k \phi_k \, dV \, , \qquad (3.1)$$

where ρ_k, ϕ_k and $\boldsymbol{J}_k$ are the density, the body source and the flux of phase k. This means the rate of change of $\rho_k \psi_k$ in V is determined by the flux coming through the surface and the generation of ψ_k through the source ϕ_k. Using the Reynolds transport theorem and

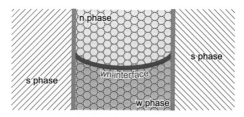

Figure (3.1): Schematic illustration of two–phase flow through a capillary, in 2D constrained by solid walls on left and right side. Further important quantities are the curvature κ of the wn–interface and the contact angle α as its boundary condition with respect to the solid wall.

the divergence theorem this results in the general Eulerian conservation law

$$\frac{\partial \rho_k \psi_k}{\partial t} + \nabla \cdot (\boldsymbol{v}_k \rho_k \psi_k) = -\nabla \cdot \boldsymbol{J}_k + \rho_k \phi_k \,. \tag{3.2}$$

Considering the momentum balance, the equation of motion can be derived with

$$\psi_k = \boldsymbol{v}_k \tag{3.3}$$

$$\boldsymbol{J}_k = -\boldsymbol{T}_k = p_k \boldsymbol{I} - \boldsymbol{\tau}_k \tag{3.4}$$

$$\phi_k = \boldsymbol{g}_k \,, \tag{3.5}$$

which results again in Lagrangian formulation in

$$\rho_k \frac{D\boldsymbol{v}_k}{Dt} = -\nabla p_k + \nabla \cdot \boldsymbol{\tau}_k + \rho_k \boldsymbol{g}_k \,. \tag{3.6}$$

Here $\boldsymbol{\tau}_k$ is the deviatoric stress tensor of phase k, p_k its pressure and $\boldsymbol{g}_k$ the gravitational acceleration.

For incompressible Newtonian fluids the deviatoric stress tensor is given by

$$\boldsymbol{\tau}_k = \mu_k \left(\nabla \boldsymbol{v} + \nabla \boldsymbol{v}^t\right) \tag{3.7}$$

with the dynamic viscosity μ_k of phase k. Therewith, the Navier–Stokes equations for incompressible Newtonian fluids under isothermal conditions are given in Lagrangian description by the following momentum balance and mass balance

$$\rho_k \frac{D\boldsymbol{v}_k}{Dt} = -\nabla p_k + \mu_k \Delta \boldsymbol{v}_k + \rho_k \boldsymbol{g} \,, \tag{3.8}$$

$$\frac{1}{\rho_k} \frac{D\rho_k}{Dt} = \nabla \cdot \boldsymbol{v}_k = 0 \,. \tag{3.9}$$

From a physical point of view, the phases are connected at interfaces and they interact through this interface with each other. Therefore, appropriate interface conditions have to be identified, which mathematically couple the equations of motion and are in general responsible for a physical mass, momentum and energy exchange through the interfaces. Note that for immiscible fluids under isothermal conditions there is only a interfacial momentum balance as the one and only interfacial condition.

3.1 Two–phase interface condition

As soon as two phases are considered in the system, an interfacial balance equation is necessary, which is responsible for the two–phase interaction in the theoretical model. Therefore let's first consider a "Gedankenexperiment" according to [deG03]. Assume a spherical drop of oil in water with radius R. By considering the system at rest, the energetic state changes through a slight change in the radius dR by the necessary work

$$\delta W = -p_o dV_o - p_w dV_w + \sigma dA\,, \tag{3.10}$$

where p_o, p_w and σ are the pressure of oil, water and the surface tension coefficient. Using $dV_o = 4\pi R^2 dR = -dV_w$ and $dA = 8\pi R dR$, the Young–Laplace equation for mechanical equilibrium $\delta W = 0$ is obtained

$$\Delta p_{ow} = p_o - p_w = 2\sigma/R\,. \tag{3.11}$$

This can be generalized to

$$\Delta p_{wn} = \sigma_{wn}\kappa_{wn}\,, \tag{3.12}$$

which provides an interface condition for a two–phase system at rest. The indices w, n and later s denote the wetting, the non-wetting and the solid phase. A combination of any two of them, e.g. wn, labels an interfacial property like p_{wn}, which is the pressure jump across the interface $p_{wn} = p_w - p_n$. σ_{wn} and κ_{wn} denote the surface tension coefficient and the curvature of the wn-interface. The latter is defined by

$$\kappa_{wn} = -\nabla \cdot \hat{\boldsymbol{n}}_{wn} \tag{3.13}$$

with $\hat{\boldsymbol{n}}_{wn}$ being the unit normal $\hat{\boldsymbol{n}} = \boldsymbol{n}/|\boldsymbol{n}|$ of the interface.

For a system under motion the resulting vectorial interface condition is given by [Lan87]

$$(p_w - p_n)\,\hat{\boldsymbol{n}}_{wn} = (\boldsymbol{\tau}_w - \boldsymbol{\tau}_n)\hat{\boldsymbol{n}}_{wn} - \sigma_{wn}\kappa_{wn}\hat{\boldsymbol{n}}_{wn} + \nabla\sigma_{wn}\,. \tag{3.14}$$

In this way the previous formulation is extended by the balance of viscous stresses and an interfacial stress gradient, which is oriented tangential to the interface. In this work the surface tension coefficient σ_{wn} is supposed to be constant along the interface, which leads to $\nabla \sigma_{wn} = 0$.

At first glance, there are two ways how the interface condition can be realized. One way is to describe the interfacial interaction by the coupling condition right between these two nodes, which are separated by the interface. This means a sharp interface is assumed to be right between the interfacial nodes, which is quite challenging for curved interfaces in $3D$, especially for smooth numerical descriptions, which don't allow for a sharp interface description by nature. The other way is to impose the interface condition in a smooth formulation on the existing nodes in the vicinity of the interface in addition to the discrete formulation of the governing equations they already obey. This means the one and only term, which has to be incorporated, is the second term on the right–hand side of the interface balance equation (3.14). This was first described by Brackbill et al. [Bra92] and he called this term the Continuum Surface Force (CSF)

$$\boldsymbol{f}_{wn} = \sigma_{wn}\kappa_{wn}\hat{\boldsymbol{n}}_{wn} \, . \tag{3.15}$$

This means the surface tension force is always directed tangential to the interface normal. The remaining question is, how are the nodes in the vicinity of the interface identified and how is $\boldsymbol{f}_{wn}$ distributed on them. The idea is to find a distribution function, which is zero inside the bulk phase and deviates from zero, if and only if the node is located in the vicinity of an interface. Furthermore, the interface condition (3.14) is written with respect to the units on a force per area basis. Therefore the continuum surface force (3.15) has to be reformulated to a force per volume basis in order to incorporate it into the momentum balance. Hence, a volume reformulation is searched in a way that

$$\boldsymbol{F}_{wn}^{vol} = \boldsymbol{f}_{wn}\delta_{wn} \, , \tag{3.16}$$

where δ_{wn} makes the transformation and represents the desired distribution function. This function is on the one hand responsible to identify the nodes close at an interface and on the other hand takes care of the volume reformulation, which means the dimension has to be $1/m$. Moreover, it has to be normalized through

$$\int_{wn} \delta_{wn}dn_{wn} = 1 \tag{3.17}$$

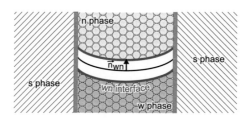

Figure (3.2): Illustration of broadened, smooth interface area.

to preserve the integral quantity of the continuum surface force and in the same way ensure the appropriate unit. Here dn_{wn} denotes the line integral across the wn-interface. This means in order to introduce a smooth transition area from the wetting phase to the non–wetting phase, cf. Fig. 3.2, a so–called color function is introduced, which makes a step by the value of 1 from one phase to the other, cf. Fig. 3.3(a). First the Heaviside step function is used for this purpose with a sharp phase transition at $r' = 0$ by a color value $c = 0$ for the phase located at $r' < 0$ and a color value of $c = 1$ for the phase located at $r' \geq 0$, cf. Fig. 3.3(a). This function is not differentiable and is therefore convoluted with a kernel as shown in Fig. 3.3(b). The result

$$\bar{c}(\boldsymbol{r}) = \int c(\boldsymbol{r}')W(\boldsymbol{r} - \boldsymbol{r}',h)\, d^3 r' \tag{3.18}$$

is the smooth color function as illustrated in Fig. 3.3(c). Now the derivative is obtained through

$$\nabla\bar{c}(\boldsymbol{r}) = \int c(\boldsymbol{r}')\nabla W(\boldsymbol{r} - \boldsymbol{r}',h)\, d^3 r' \tag{3.19}$$

and visualized in Fig. 3.3d. Note the applied trick in this context. By introducing an antiderivative with a predefined jump at the considered interface it is ensured, that the integral of the derivative equals the predefined value of the jump in the antiderivative. In this way the transformation δ_{wn} is constructed, which satisfies Eq. (3.17). This procedure provides yet another advantage. As the gradient is always directed normal to the interface, an interface normal can be determined by

$$\boldsymbol{n}_{wn} = \frac{\nabla\bar{c}}{[c]}, \tag{3.20}$$

where $[c]$ denotes the jump in the value of the color function (only necessary if $\neq 1$) when crossing an interface and can be considered as renormalization. Therewith the transformation and the volume reformulation is given by

$$\delta_{wn} = |\boldsymbol{n}_{wn}| \ . \tag{3.21}$$

In this manner the Navier–Stokes equations are extended by Eq. (3.16) in order to obtain

$$\frac{D\boldsymbol{v}}{Dt} = -\frac{1}{\rho}\nabla p + \frac{\mu}{\rho}\Delta\boldsymbol{v} + \boldsymbol{g} + \frac{1}{\rho}\boldsymbol{F}_{wn}^{vol} \ . \tag{3.22}$$

This formulation is capable of describing a two–phase system including the phenomena of surface tension. In this way there is no explicit formulation of the interface necessary. But still, the contact with a third solid phase cannot be handled.

Inspired by this approach a similar description for the three–phase contact area is introduced in the next chapter for the Contact Line Force, which is responsible for a dynamical description of evolving contact lines and in this way describing the wetting dynamics.

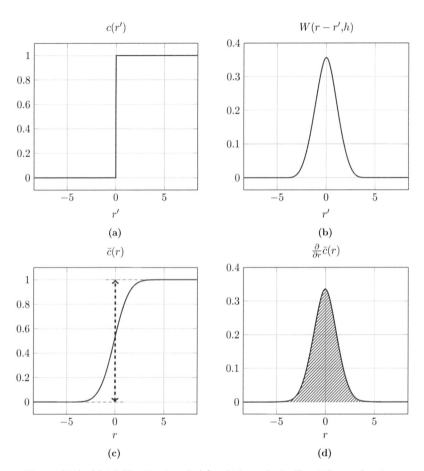

Figure (3.3): (a) $1D$ Visualization of $c(r')$, which equals the Heaviside step function as starting condition for the phase transition. (b) is the curve of the 1D Wendland kernel and (c) is the visualization of the smooth color function as obtained by the convolution of Heaviside step function and the Wendland kernel. (d) shows the derivative of the smooth color function.

3.2 Three–phase contact line condition

3.2.1 Wettability

We are subconsciously dealing with wetting effects in our daily life. For example we are used to condensed water droplets on a cold window, coatings for outdoor furniture or water–repellent textiles. But even nowadays the scientific description remains partly ambiguous.

To understand the dynamics of wetting processes one first has to strip down the phenomena and look at the region, where all three phases come together and intersect at the contact line, cf. Fig. 3.4(a). In equilibrium, the occurring forces are balanced and Young's equation is defining the static contact angle α_S through

$$\cos(\alpha_S) = \frac{\sigma_{ns} - \sigma_{ws}}{\sigma_{wn}}, \tag{3.23}$$

where σ_{ns} in units of $\left[\frac{N}{m}\right]$ represents the interfacial energy of the non–wetting/solid interface, σ_{ws} the interfacial energy of the wetting/solid interface and σ_{wn} the interfacial energy of the wetting/non–wetting interface. This equation can be derived in two ways,

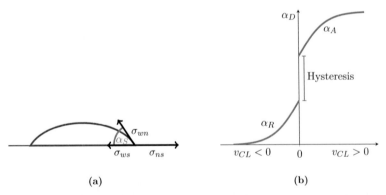

(a)   (b)

Figure (3.4): Schematic illustration of (a) balance at the three phase contact line and (b) the concept of contact angle hysteresis. By definition $v_{CL} > 0$ means wetting is taking place with an advancing contact angle α_A. For $v_{CL} < 0$ dewetting is taking place with a receding contact angle α_R. In general the relation $\alpha_R < \alpha_S < \alpha_A$ is valid.

either by using the principle of energy minimization or by applying a force balance [Bos93].
It is important to note, that this image of a sharp contact line only exists in the macroscopic
theory of this continuum–scale model. In fact the area, where all responsible processes
are taking place on the micro–scale or in principal on the molecular–scale, has a certain
extension. Within this area the distribution of fluid molecules is fluctuating around a mean
configuration [Bla93] giving rise to an image of a sharp interface with a sharp contact line
on the upscaled macro–scale.

The influence of surface roughness on the contact angle for a system at rest was already
investigated by Wenzel [Wen36]. If a structural heterogeneity is taken into account in an
energetic consideration of a contact line shift, the Wenzel condition is achieved [deG03],
giving rise to a new, effective contact angle

$$\cos \alpha_S^* = r \cos \alpha_S \,, \tag{3.24}$$

where r is a surface roughness factor, which is usually > 1 and given by the ratio of actual
surface to the geometric surface. This means hydrophilic surfaces with $\alpha_S < 90°$ become
even more hydrophilic ($\alpha_S^* < \alpha_S$) and hydrophobic surfaces with $\alpha_S > 90°$ become even
more hydrophobic ($\alpha_S^* > \alpha_S$).

Cassie and Baxter [Cas44] extended this effective consideration by taking a heterogeneous
chemical constitution of the surface into account, where each part can be characterized by
its own static contact angle. From an infinitesimal energy change due to a shift of the
contact line the Cassie–Baxter relation can be derived [deG03]

$$\cos \alpha_S^* = f^1 \cos \alpha_S^1 + f^2 \cos \alpha_S^2 \,, \tag{3.25}$$

where f^i denotes the surface fraction of species i with $i = [1,2]$ and $\sum_i f^i = 1$. Note
that this model is also valid e.g. if one species is represented by a solid state matrix and
the other species by entrapped air. Under the assumption, that the dimensions of these
heterogeneities are very small compared to the droplet size, both these conditions lead to
a new effective contact angle in the "far field" and they are applicable to systems at rest.
Taking advantage of aforementioned descriptions one is able to invent superhydrophobic
materials by design.

In this work dynamic flow phenomena are studied and therefore dynamic effects on the
contact angle and the corresponding fluid shape are investigated. Wetting is the defined

process, where a wetting phase displaces a non–wetting phase on a solid substrate. Consequently the contact line has to move towards the non–wetted area. Dewetting happens, as soon as the process is going the other way round.

In industrial applications, e.g. coating processes or petroleum recovery, the speed of the contact line motion plays a crucial role. If the motion of the bulk wetting fluid is faster than the motion of the contact line, the wetting phase may entrap parts of the fluid, which are supposed to be displaced. In this way the quality of the product or in general the outcome is affected [Bla79].

As the dynamic experiments from Hoffman [Hof75] show the contact angle may attain different values based on the velocity of the system. In a wetting case the dynamic contact angle (DCA) becomes bigger than the static contact angle $\alpha_D > \alpha_S$ and the notation advancing contact angle α_A is used. In the dewetting case with $\alpha_D < \alpha_S$ the notation receding contact angle α_R is applied. An idealized scheme how the dynamic contact angle α_D changes under motion is shown Fig. 3.4b. Note that one should distinguish between a velocity dependent contact angle and the origin of contact angle hysteresis [Sch85]. Some publications do not clearly differentiate, e.g. [Bla93, Osi62]. They already refer to contact angle hysteresis in presence of different advancing and receding contact angles, whereas [Bos93] clearly points out: the dynamic contact angle of course depends on the velocity of the system, but the indication to call this behavior a hysteretic one is given by the gap, as shown in Fig. 3.4b. In this work the wording contact angle hysteresis is only used in connection with the existence of such a gap. The origin thereof are static conditions like surface roughness and impurities or in general strucural and chemical heterogeneities [Bla93, Bos93]. With respect to wettability the consequences of these heterogeneous surface conditions are metastable configurations [Joh93, Bla93]. An exemplary illustration is given

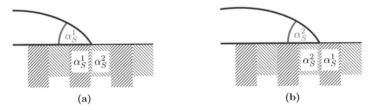

(a)   (b)

Figure (3.5): Visualization of (a) a metastable and (b) an unstable system configuration due to a surface heterogeneity with $\alpha_S^1 < \alpha_S^2$.

in Fig. 3.5. The initial state shows a solid, which has an alternating surface condition in terms of the interfacial energy of the liquid–solid interface. The scale of this heterogeneity has to be small compared to the considered droplet region and is here visualized by the two different static contact angles α_S^1 and α_S^2. For explanation $\alpha_S^1 < \alpha_S^2$ is used. The critical states are given in case of the periphery sitting right on such a boundary [Joh64]. (a) shows the equilibrium state where the contact line and the corresponding contact angle do not yet experience any influence by the heterogeneity of the solid. Now imagine to move the contact line by an infinitesimal displacement to the right. This would cause the contact angle to pass into a transitional state and assume the new equilibrium state when crossing the boundary. Due to the wetting condition there is a counterforce, which pushes the contact line back towards the wetting zone with the lower contact angle α_S^1. This is why this state is called a metastable state. (b) represents an unstable system state, because an infinitesimal small displacement of the contact line to the right would destroy the whole configuration.

The variety of these possible contact angle states gives rise to contact angle hysteresis showing different, velocity dependent dynamic contact angles.

Literature and experimental results are very controversial regarding the existence of upper and lower limits for advancing and receding contact angles [Bla93, Bla02, Lam02]. It is perceived that the type of fluid viscosity also has an influence on these limits and on the dependency of the contact angle with respect to the flow velocity [Kim15]. Additionally, it is found that, there are different wetting regimes based on surface wettability and liquid viscosity. For example if the contact angle is below a certain threshold value, viscosity becomes dominant in the wetting process [Che14].

Different models for dynamic contact angles were developed in the past. Blake and Haynes built a constitutive relationship based on the molecular–kinetic theory [Bla69]. In this model adsorption and desorption rates of fluids on solid substrates are incorporated to describe a progressing contact line. De Gennes and Brochard–Wyart suggested a hydrodynamical approach [deG85, Bro92] in order to obtain a dependency of the dynamic contact angle on the velocity of the contact line. They assumed a force balance within a region around the moving contact line by the "non compensated Young Force" and the viscous friction of a parabolic Poiseuille flow profile, which originates from the no–slip condition at the wall. Shikhmurzaev introduced a similar model for gas–liquid interfaces

[Shi94, Shi96] and for liquid–liquid interfaces [Shi97] by deriving the governing equations from first principles with an explicit treatment of the interface and the contact line area. The above mentioned models were introduced to obtain constitutive relations of the dynamic contact angle and the contact line velocity. In this way they have the assumption of a steady state motion in common and there are many experiments proofing each one of it. A comparison of these different methods is therefore beyond the scope of this work. Instead the basic idea is utilized to develop a physico–numerical model, which is capable of describing two–phase flow in capillaries and droplet dynamics respectively with surface tension and contact line motion.

Motivated by the idea of Brochard–Wyart and de Gennes [Bro92] to superimpose the viscosity model in a region close to a wall with the unbalanced Young force, in this work also a comparable approach is introduced, now for a spatially resolved model. It is worth mentioning, that in the macroscopical consideration Shikhmurzaev [Shi97] also ended up with the same dependency of the total drag force before he proceeded to the constitutive relation. An additional advantage of this approach is, that in the same way as Brackbill et al. [Bra92] integrated the momentum balance of a two–phase interface to obtain the Continuum Surface Force (CSF) now the momentum balance of a three–phase contact line is taken into account to obtain the Contact Line Force (CLF). In this way a slip model in the immediate vicinity of the contact line is introduced. By leaving the vicinity of the contact line, a transition to the no–slip area is intrinsically realized through an adapted volume reformulation, following the principle of Brackbill et al. [Bra92]. In this manner the singularity at the contact line due to the no–slip condition and the requested contact line motion is eliminated, which is a necessary step according to the comment of Blake [Bla93]:

> "For example, as a result of hydrodynamic theoretical studies, it is now widely recognized that it is necessary to relax the classical boundary condition of no slip at the wall and allow limited slip between fluid and solid in the immediate vicinity of a moving wetting line."

3.2.2 State of the art models

In principle all models here have to get rid of the singularity at the contact line. That's not how they are distinguished in this work. In fact they are categorized by the physical model, which describes the dynamical wetting in terms of a model for the dynamic contact angle. This can have a strong influence on the dynamics, as the contact angle is the boundary condition to the curvature of the fluid–fluid interface. By now, there are in principal four different physico–numerical methods to describe the local wetting behavior on the continuum scale:

Static contact angles (and dynamic contact angles with constitutive equations)

Models in this category use a static contact angle or an in advance defined dynamic contact angle and require a separate description for interfaces and contact lines. Usually the CSF model is used to describe surface tension phenomena of fluid–fluid interactions and different models for contact angles are possible. Using the "Moving Particle Semi–implicit (MPS)" method, Liu et al. [Liu05] simulated drop shapes and the wettability of walls with static contact angles. Ferrari and Lunati [Fer13b] simulated two–phase drainage processes using static contact angles and the VOF method in porous media. Annapragada et al. [RA12] simulated drop shapes on an incline with a static boundary condition for the contact angle and a constitutive equation for the azimuthal angle of the drop. Das and Das [Das10] also use a constitutive equation to obtain volume–dependent contact angles for bigger drops. Approaches with static contact angles are sufficient for stationary problems as well as dynamical processes with low capillary numbers.

Also models with constitutive equations for "dynamic" contact angles in dependence of a contact line velocity are included in this category, e.g. Sikalo et al. [Sik05]. For a specific state of the system the dynamic contact angle therewith is an input parameter and not a result of the model. This means, static contact angles and advancing or receding contact angles with surface tension coefficients are input parameters in these models.

A way to realize the motion under static, given contact angles is to include it in the slip model. For a deeper insight the work of Ren and E [Ren07] is referred to. They developed a slip model with boundary conditions for the moving contact line problem and introduced an effective friction coefficient for the slip model.

Breinlinger et al. [Bre13] published a SPH approach, which imposes static contact angles at the wall. In order to obtain static contact angles in equilibrium, the underlying idea is to introduce a reset force through an artificially generated curvature. This way a correction of normals close to the wall and an additional smoothing has to be introduced. Due to the layer dependent formulation in this model, there might rise a peak curvature, which exerts an unphysical force. This is why the force is an artificial reset force without the claim for correct process dynamics, as there is no momentum balance used to obtain this force.

Pairwise interaction forces

Particle–particle interactions can be introduced to model the behavior of interfaces and contact lines unified in one interaction following the basic idea of molecular dynamics and Van der Waals forces. Tartakovsky and Meakin [Tar05] were the first, who published an effective formulation of surface tension with inherent contact angles in SPH. Zhou et al. [Zho08] utilized a Lennard–Jones–like potential with adapted coefficients for continuum scale consideration. An implementation in the Volume of Fluid (VOF) method is published by Mahady et al. [Mah15]. This shows the general ability to choose an artificial interaction potential and use an appropriate fitting parameter to model the qualitative behavior of surface tension phenomena. Beside the classical fitting approach, there is now also a derivation available, which shows how coefficients for pairwise interaction can be determined, see Bandara et al. [Ban13].

Stress tensor formulations

A multi–phase model in terms of a stress tensor formulation was introduced by Lafaurie et al. [Laf94]. The result is a momentum conserving propagation of interfaces and contact lines unified in one description. Hu and Adams [Hu06] showed an implementation of this method in SPH. The input parameters for this model are the surface tension coefficients of all involved interfaces.

Within the limiting factor by the applied particle resolution and smoothing length, this approach uniformly distributes the influence by the stress tensor formulation in a geometrical sense around the interface. With respect to forces this might sound reasonable. However, acceleration is force per mass, which means at an air–water interface with a high density ratio the acceleration of the air phase may be much higher and reach a ratio of 1000:1 compared to the water phase [Ada10]. This dramatically decreases the time step size due to the time step criterion and in practice has a strong influence on the stability of the interface and in this way on the whole considered system. Moreover, two smoothings are applied in a row, first the divergence of the color function and afterwards the divergence of the stress tensor. This means the influence area around an interface is enlarged.

Diffuse interface method

Yet another way to get rid of the singularity is the Diffuse Interface method (DI) as published by [Jac00]. In this method also a smooth transition is introduced, which is in general an appropriate way for non–smoothing methods. This means it is not mandatory to introduce an additional smoothing in SPH, because the method comes with an inherent smoothing. But if necessary, the DI method can be used in SPH to decouple an additional, artificially introduced smoothing of an interface from the numerical, intrinsic smoothing. An application to SPH is published by Das and Das [Das10, Das11].

At low capillary and Bond numbers, surface tension is the dominating force in the momentum balance which justifies the assumption of static contact angles in these flow regimes. At high capillary numbers, viscous flow phenomena and body forces become dominating. In these flow regimes it may be possible to neglect surface tension and still obtain acceptable results depending on the desired use case. But in the transition regime the negligence of surface tension and therewith dynamic contact angles may lead to bad results. To model dynamic contact angle as a result of the simulation the contact line force model by Huber et al. [Hub16b] is now introduced.

The big advantage of this approach is, that one can easily exchange the driving force if other wetting conditions are considered and moreover, it is adjustable to high density ratios.

3.2.3 Contact Line Force (CLF)

Balance equations can be formulated for phases, interfaces and contact lines, as Hassanizadeh and Gray [Has93] showed. Starting from the momentum balance of a contact line the Navier–Stokes equations can be extended in the same way for the contact line as previously explained in chapter 3.1 for an interface.

Momentum balance

Analogous to the derivation of the continuum surface force now the balance equation of the contact line is used to include the dynamical behavior with dynamic contact angles. Starting from the *Microscale Momentum Balance Equation for a Contact Line* by Hassanizadeh and Gray [Has93], the *unbalanced Young Force* as proposed by de Gennes

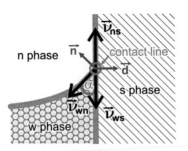

Figure (3.6): Sketch of the contact line region with contact angle α.

[deG85] can be derived. To achieve this, first the net force of all contributions due to surface tension at the contact line is considered according to Fig. 3.4(a)

$$\boldsymbol{f}_{wns}^{net} = \sigma_{wn}\hat{\boldsymbol{\nu}}_{wn} + \sigma_{ns}\hat{\boldsymbol{\nu}}_{ns} + \sigma_{ws}\hat{\boldsymbol{\nu}}_{ws} = \sum_{ij=wn,ns,ws} \sigma_{ij}\hat{\boldsymbol{\nu}}_{ij} \,. \tag{3.26}$$

Note that the direction of the surface tension coefficients is now described by the unit vectors $\hat{\boldsymbol{\nu}}_{ij}$ as illustrated in Fig. 3.6. Here $\hat{\boldsymbol{\nu}}_{ns}$ is the unit vector in–plane of the ns–interface, perpendicular on the contact line itself and pointing away from the contact line. The same description holds for $\hat{\boldsymbol{\nu}}_{ws}$ and $\hat{\boldsymbol{\nu}}_{wn}$. A further step towards an even more detailed model could be the introduction of a contact line tension in terms of [Ami00], which is not considered in this approach here so far. The formulation of Eq. (3.26) can be decomposed into linearly independent, orthogonal parts. Two directions are now examined more in detail. First the normal direction with respect to the solid wall, described by the unit vector $\hat{\boldsymbol{d}}$, and later, the in–plane component of the fluid–solid interface, described by the unit vector $\hat{\boldsymbol{\nu}}_{ns}$. The projection in normal direction towards the solid wall is given by

$$\boldsymbol{f}_{wns}^{\perp} = \hat{\boldsymbol{d}}\,\hat{\boldsymbol{d}} \cdot \sum_{ij=wn,ns,ws} \sigma_{ij}\hat{\boldsymbol{\nu}}_{ij} \,. \tag{3.27}$$

The inner product of $\hat{\boldsymbol{d}}$ and a surface tension component in–plane of the fluid–solid interface $\hat{\boldsymbol{\nu}}_{is}$ vanishes due to orthogonality. In this way the remaining contribution is

$$\boldsymbol{f}_{wns}^{\perp} = \sigma_{wn}(\hat{\boldsymbol{d}} \cdot \hat{\boldsymbol{\nu}}_{wn})\,\hat{\boldsymbol{d}} = -\sigma_{wn}\sin(\alpha_D)\,\hat{\boldsymbol{d}} \,. \tag{3.28}$$

In order to distinguish the actual, dynamic contact angle from the static contact angle, the denotations α_D and α_S are introduced. When the system is at rest, α_D becomes α_S. In this work all walls are assumed to be fixed, rigid bodies, which means there is no net force left in normal direction of the wall or with other words the normal components are in balance. Therefore $\boldsymbol{f}_{wns}^{\perp}$ does not have to be implemented in the simulation and the only contribution of surface tension at the contact line is given by the parallel component

$$\boldsymbol{f}_{wns}^{\parallel} = \hat{\boldsymbol{\nu}}_{ns}\,\hat{\boldsymbol{\nu}}_{ns} \cdot \sum_{ij=wn,ns,ws} \sigma_{ij}\hat{\boldsymbol{\nu}}_{ij}\,. \tag{3.29}$$

In this way a force acting on the contact line is introduced by only considering tangential components of surface tension. For convenience now the denotation $\boldsymbol{f}_{wns} = \boldsymbol{f}_{wns}^{\parallel}$ is used, implying that only tangential contributions are considered and necessary for the dynamic propagation of the system. As $\hat{\boldsymbol{\nu}}_{ns}$ and $\hat{\boldsymbol{\nu}}_{ws}$ are oppositely directed, the equivalent formulation $\hat{\boldsymbol{\nu}}_{ws} = -\hat{\boldsymbol{\nu}}_{ns}$ is used in the following step

$$\boldsymbol{f}_{wns} = [\sigma_{wn}\hat{\boldsymbol{\nu}}_{wn} + \sigma_{ns}\hat{\boldsymbol{\nu}}_{ns} + \sigma_{ws}\hat{\boldsymbol{\nu}}_{ws}] \cdot \hat{\boldsymbol{\nu}}_{ns}\,\hat{\boldsymbol{\nu}}_{ns} \tag{3.30}$$

$$= [\sigma_{wn}\hat{\boldsymbol{\nu}}_{wn} \cdot \hat{\boldsymbol{\nu}}_{ns} + \sigma_{ns} - \sigma_{ws}]\,\hat{\boldsymbol{\nu}}_{ns} \tag{3.31}$$

$$= [-\sigma_{wn}\cos(\alpha_D) + \sigma_{ns} - \sigma_{ws}]\,\hat{\boldsymbol{\nu}}_{ns}\,. \tag{3.32}$$

From this formulation Young's equation can be derived. By considering the system at rest, the previous equation is balanced, giving rise to

$$0 = -\sigma_{wn}\cos(\alpha_S) + \sigma_{ns} - \sigma_{ws}\,, \tag{3.33}$$

which exactly is Young's equation as introduced in Eq. (3.23)

$$\cos(\alpha_S) = \frac{\sigma_{ns} - \sigma_{ws}}{\sigma_{wn}}\,.$$

Using this relation, the substitution $\sigma_{ns} - \sigma_{ws} = \sigma_{wn}\cos(\alpha_S)$ is inserted in Eq. (3.32) in order to obtain

$$\boldsymbol{f}_{wns} = [-\sigma_{wn}\cos(\alpha_D) + \sigma_{wn}\cos(\alpha_S)]\,\hat{\boldsymbol{\nu}}_{ns} \tag{3.34}$$

$$= \sigma_{wn}\,[\cos(\alpha_S) - \cos(\alpha_D)]\,\hat{\boldsymbol{\nu}}_{ns}\,. \tag{3.35}$$

This means in a general unbalanced case the necessary input parameters for this Contact Line Force model are either all three surface tension coefficients or by knowing the static contact angle, it's sufficient to additionally know just the surface tension coefficient of the fluid–fluid interface.

Volume reformulation

As previously explained, the momentum balance equation for an interface is formulated as force per area. The principle of the volume formulation for an interface is now adapted for the description of contact lines. In this case the momentum balance is formulated as force per line, hence the volume reformulation for a contact line

$$\boldsymbol{F}^{vol}_{wns} = \boldsymbol{f}_{wns}\delta_{wns} \tag{3.36}$$

is on the one hand responsible to transform the force per line to a force per volume and on the other hand responsible for distributing the forces, which act on the contact line, to the nodes in the vicinity of the line.

In this way the transformation has to satisfy

$$\iint \delta_{wns}\, dA = 1 \quad \text{with} \tag{3.37}$$
$$dA = |\boldsymbol{d} \times \boldsymbol{\nu}_{ns}|\, dd\, d\nu_{ns}\,,$$

which is a two–dimensional integration in the plane orthogonal to the contact line. The basis vectors, which are chosen here for the integration, are the tangential vector $\boldsymbol{\nu}_{ns}$ and the normal vector $\boldsymbol{d}$. In principle various basis vectors could be used. For simplicity they just should be perpendicular to each other in order to build a cartesian system. The benefit in choosing these two basis vectors is, that by commiting to the last one, one can utilize a discrete representation of the wall for an evaluation in direction normal to the wall. And the choice of the last one determines the usage of the first vector $\boldsymbol{\nu}_{ns}$.

Now the same trick is applied as explained in the volume reformulation for an interface. In order to obtain a distribution, which satisfies Eq. (3.37), an arbitrary antiderivative is searched, which has a smooth jump by a predefined value. This function is already known. Again, the color function can be utilized, but to obtain the appropriate transformation δ_{wns}, now two derivatives must be applied in a row. Therefore δ_{wns} is defined in the following way. In a first step, the directional derivative is taken in the direction of the

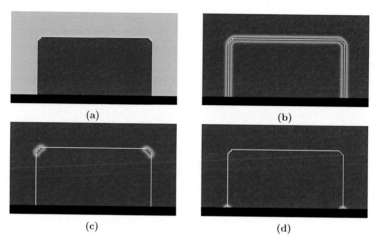

Figure (3.7): Initial state, (a) colors represent different phases: solid wall in black, wetting phase in grey and non–wetting phase in blue. (b) Magnitude of fluid interface normal $|\boldsymbol{n}_{wn}| = \delta_{wn}$, (c) curvature κ and (d) volume reformulation of CLF: δ_{wns}. Color scale from blue (zero value) to red with increasing magnitude.

fluid–solid interface, as defined through

$$\delta'_{wns}(\boldsymbol{r}) = \frac{1}{[c]} \nabla_{\boldsymbol{\nu}_{ns}} \bar{c} = \hat{\boldsymbol{\nu}}_{ns} \cdot \boldsymbol{n}_{wn} \tag{3.38}$$

and only evaluated on the fluid phase. $\nabla_{\boldsymbol{\nu}_{ns}}$ is the gradient along a direction with

$$\nabla_{\boldsymbol{\nu}_{ns}} = \hat{\boldsymbol{\nu}}_{ns} \cdot \nabla . \tag{3.39}$$

Note that apart from the directional derivative this is the same procedure as for the volume reformulation of an interface. In case of the continuum surface force, the magnitude of the interface normal is used for the volume reformulation $\delta_{wn} = |\boldsymbol{n}_{wn}|$, cf. Fig. 3.7(b). Here only the inner product with the tangential direction and the interface normal has to be applied in addition.

In a second step, a further derivative is built, orthogonal to the previous one, in direction normal to the solid boundary $\hat{\boldsymbol{d}}$ by

$$\delta_{wns}\left(\boldsymbol{r}\right) = -2\,\nabla_{\boldsymbol{d}}\delta'_{wns}\,. \tag{3.40}$$

In this step, a discrete boundary is taken into account with $\delta'_{wns} = 0$ and the derivative is solely taken with respect to the boundary and evaluated on fluid elements but not among them. This gives rise to a well–defined transition right at the boundary, cf. Fig. 3.7(d). The factor 2 is introduced, because in this step only the influence on the fluid phase is evaluated and this represents only half of the kernel support at the contact line. Due to the fact that δ'_{wns} is positive inside the fluid domain, the directional derivative towards the solid wall becomes negative and the minus sign is introduced to compensate this in order to obtain a positive weighting $\delta_{wns} > 0$. In this way δ_{wns} is used for the volume reformulation of the contact line force. In the same way this description is responsible for distributing the contact line force to the nodes close to the contact line. In this manner the quantity of the contact line force is preserved, which can be checked by integrating δ_{wns} in the fluid domain. A validation thereof is shown in chapter 5.

With this volume reformulation the complete two–phase momentum balance is given by

$$\frac{D\boldsymbol{v}}{Dt} = -\frac{1}{\rho}\nabla p + \frac{\mu}{\rho}\Delta\boldsymbol{v} + \boldsymbol{g} + \frac{1}{\rho}\boldsymbol{F}^{vol}_{wn} + \frac{1}{\rho}\boldsymbol{F}^{vol}_{wns}\,, \tag{3.41}$$

which can be used to model two–phase flow phenomena including a complete surface tension model for the dynamics of partial wetting with dynamic contact angles. An advantage of this new formulation is, that driving forces for contact lines $\boldsymbol{f}_{wns}$ can easily be exchanged without further adjustments of the equation or the formalism, e.g. when other wetting phenomena with different wetting models are subject of new investigations.

After this introduction to the physical model, the discrete formulation thereof is explained now. As a first insight an overview of the evolution of SPH is given, which shows the roadmap from astrophysical applications to the phenomena of interest for this work. Afterwards, the actual numerical implementation of the two–phase flow model is explained more in detail including the description of boundary conditions and time discretization.

4 Present SPH model

4.1 Background

SPH came up in the late 1970s when Gingold and Monaghan were looking for a new way to model hydrodynamical equations for astrophysical phenomena [Gin77]. The idea was to introduce statistical techniques to describe analytical fields such that scattered information is projected on a continuous scale for an effective calculation. Especially in cosmic space, where lots of "empty" space exists, one does not want to waste computational power to areas, which have no influence and are not of any interest.

Lucy in principal had the same idea and started his argumentation based on Monte Carlo theory with the aim to model the evolution of protostars [Luc77]. Because of the analogy between astrophysical phenomena and fluid–dynamics other applications were found soon. A more general overview of this method for different applications reaching from viscosity modeling over heat conduction up to elasticity is given by Monaghan [Mon05]. In Rosswog [Ros09] rather fundamental physical questions regarding conservation laws and numerical challenges in terms of different discretization schemes are investigated. Randles and Libersky [Ran96] addressed the issue of boundary conditions and brought the method to applications like impacts, explosions and fractures. With respect to single–phase dynamics, Morris et al. [Mor97] showed an implementation for fluid–solid interfaces with discretized solid particles to model no–slip boundary conditions. With respect to the incompressibility model, there are in principal two implementations: the weakly compressible SPH scheme (WCSPH), which uses a stiff artificial equation of state, and the so–called truly incompressible SPH (ISPH). As in this work only ISPH is of interest, the PhD thesis of F. Keller is referred to for a deeper insight to WCSPH [Kel14]. Further comparisons of both SPH schemes can be looked up for test cases like the lid–driven cavity, the dam–break problem and the flow around an obstacle in [Lee08, Che13, Sef15]. The last one also includes an example of the the Karman vortex.

As soon as a second phase is involved, plenty academic test cases are available and necessary for verification of the code. Therefore classical two–phase flow instabilities are usually subject of comprehensive investigations like Rayleigh–Taylor instabilities, see [Cum99, Hu07, Sze12b, Mon13] and Kelvin–Helmholtz instabilities [Pri08, Val10, Rea10, Fat14], respectively. Other test cases deal e.g. with rising bubbles [Col03, Gre09, Gre13, Sze15], the coalescence collision of liquid drops [AM11b, AM11a] or Taylor–Green vortices [Hu07, Ada13] etc. .

As there are many applications, which have to deal with high density ratios, a stable model under these circumstances is of huge interest [Hu06, Ada10]. An interesting unified implementation of surface tension including the description of contact angles is achieved by the stress tensor formulation, as shown by Hu and Adams [Hu06]. In this way Grenier et al. [Gre09] showed a SPH–formulation to describe real water–gas interfaces in terms of bubbly flows. This method is also referred to as Continuous Surface Stress (CSS) method [Gre13].

Especially with respect to high density ratios for bubbles and droplets, a separate description of interface dynamics and contact line dynamics is realized here in this work. Therefore the well–established SPH formulation of the Continuum Surface Force model by Morris [Mor00] is used and the adjusted description for high density ratios by Adami et al. [Ada10] is utilized. On top of this, the new formulation for contact line dynamics, as introduced by Huber et al. [Hub13, Hub16b, Hub16a], is applied. This Contact Line Force model completes the interaction of the Continuum Surface Force, when two–phase interfaces approach a third, solid phase. The numerical model is now built one by one.

4.2 Navier–Stokes equations

The common discretization schemes were introduced in chapter 2.4. In this work basically the models of Adami et al. [Ada10] and Szewc et al. [Sze12a] are utilized. At first glance, the latter viscosity model is interesting, because Eq. (2.92) conserves linear and angular momentum. This is why this model is investigated first. The viscosity term of the momentum balance is given by

$$
(\nabla(\eta\nabla \cdot \boldsymbol{v}))_a = \sum_b 8m_b \underbrace{\frac{\eta_a + \eta_b}{\rho_a + \rho_b} \frac{1}{|\boldsymbol{r}_{ab}|^2 + \zeta^2}}_{\xi_{ab}} \boldsymbol{v}_{ab} \cdot \boldsymbol{r}_{ab} \, \widetilde{\nabla}_a \widetilde{W}_{ab} \,,
\tag{4.1}
$$

where the corrected SPH formulation of Eq. (2.39) is from now on applied for all derivatives. Consequently the pressure term in the momentum balance is discretized by

$$\left(\frac{1}{\rho}\nabla p\right)_a = \sum_b m_b \frac{p_a + p_b}{\rho_a \rho_b} \widetilde{\nabla}_a \widetilde{W}_{ab}. \qquad (4.2)$$

With respect to mass conservation, the multi–phase formulation for the density by Hu and Adams [Hu06] is used

$$\rho_a = m_a \sum_b \widetilde{W}_{ab}. \qquad (4.3)$$

The projection method of Cummins and Rudman [Cum99] leads to the discrete formulation of the pressure Poisson equation, cf. Eq. (2.76), with a left–hand side of

$$\left(\nabla \cdot \left(\frac{1}{\rho}\nabla p\right)\right)_a = \sum_b V_b \underbrace{\frac{4}{\rho_a + \rho_b} \frac{\boldsymbol{r}_{ab} \cdot \widetilde{\nabla}_a \widetilde{W}_{ab}}{|\boldsymbol{r}_{ab}|^2 + \zeta^2}}_{\chi_{ab}} p_{ab} \qquad (4.4)$$

in the linear equation system. The right–hand side is described by

$$\frac{1}{\Delta t}\left(\nabla \cdot \boldsymbol{v}_{int}\right)_a = \frac{1}{\Delta t}\sum_b V_b \left(\boldsymbol{v}_{int_b} - \boldsymbol{v}_{int_a}\right) \cdot \widetilde{\nabla}_a \widetilde{W}_{ab}. \qquad (4.5)$$

Note that due to the corrective scheme for the density, cf. Eq. (4.3), the right–hand side with a formulation of $\nabla \cdot \boldsymbol{v}$ can't be replaced through the continuity equation by a discrete formulation of $-\frac{1}{\rho}\frac{\partial \rho}{\partial t}$ anymore, because a field of constant mass particles is through the zeroth order consistent formulation always constant. Consequently the density field in this model is decoupled from the particle arrangement.

4.3 Surface tension

As only the wn–interface is of interest, for readability reasons the labels of interface normal and curvature of particle a are now reduced from $\boldsymbol{n}_{wn_a}$ to $\boldsymbol{n}_a$ and accordingly from κ_{wn_a} to κ_a. In the same way $\boldsymbol{\nu}_{ns_a}$ is reduced to $\boldsymbol{\nu}_a$. The interface normal is evaluated through

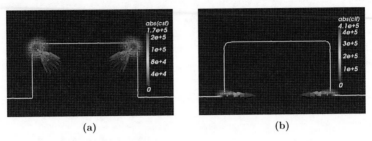

(a)   (b)

Figure (4.1): Initial state: (a) Visualization of the Continuum Surface Force (CSF). (b) Visualization of the Contact Line Force (CLF). Static contact angle is set to $140°$.

the jump in the color function according to Eq. (3.20) and results in

$$\boldsymbol{n}_a = \frac{1}{[c]} \sum_b V_b \left(c_b - c_a\right) \widetilde{\nabla}_a \widetilde{W}_{ab} .\tag{4.6}$$

In the next step the unit normal is used to calculate the curvature κ through

$$\kappa_a = -\sum_b V_b \left(\hat{\boldsymbol{n}}_b - \hat{\boldsymbol{n}}_a\right) \cdot \widetilde{\nabla}_a \widetilde{W}_{ab} ,\tag{4.7}$$

where the summation is applied only to particles with $|\boldsymbol{n}| > 0.01/h$, cf. [Mor00], because otherwise the particles sitting far away from the interface (while still being close enough to obtain an interface normal) would strongly contribute and lead to a worse result of the curvature. Note that if the "+"–formulation of Eq. (2.25) would be used for the calculation of the interface normal in Eq. (4.6) and no corrective approach is applied, such a filter would be necessary to get rid of the background noise in $|\boldsymbol{n}_a|$, which is caused by the particle anisotropy and the non–zeroth order consistent formulation.

The surface force then writes as

$$\boldsymbol{F}_{wn_a}^{vol} = \sigma_{wn} \kappa_a \boldsymbol{n}_a .\tag{4.8}$$

The result is shown in Fig. 4.1(a). Plane interfaces don't cause surface forces and the stronger the curvature the stronger the continuum surface force.

To determine the dynamic contact angle α_D, the distance vector

$$\boldsymbol{d}_a = \sum_b V_b \, \boldsymbol{r}_{ba} \, \widetilde{W}_{ab} \tag{4.9}$$

is used, which points straight to the wall (cf. Fig. 3.6), where summation is applied over solid boundary particles only. At domain boundaries also the ghost particles could be used instead. But that's anticipated here and will be explained in the next section. The contact angle is obtained by

$$\cos(\alpha) = \hat{\boldsymbol{d}}_a \cdot \hat{\boldsymbol{n}}_a \,, \tag{4.10}$$

which always gives rise to the contact angle of that phase, the interface normal $\hat{\boldsymbol{n}}_a$ is pointing to. The interface normal is by definition always pointing towards the phase with the higher color value c. This means in Fig. 3.6 the wetting phase has the lower color value and therefore the dynamic contact angle of that phase is given by

$$\cos(\alpha_D) = -\hat{\boldsymbol{d}}_a \cdot \hat{\boldsymbol{n}}_a \,. \tag{4.11}$$

Consequently the contact line force of Eq. (3.35) or Eq. (3.32) is implemented by

$$\boldsymbol{f}_{wns_a} = \sigma_{wn} \left[cos(\alpha_S) + \hat{\boldsymbol{d}}_a \cdot \hat{\boldsymbol{n}}_a \right] \hat{\boldsymbol{\nu}}_a \tag{4.12}$$

with

$$\boldsymbol{\nu}_a = |\boldsymbol{d}_a|^2 \, \boldsymbol{n}_a - (\boldsymbol{d}_a \cdot \boldsymbol{n}_a) \, \boldsymbol{d}_a \,. \tag{4.13}$$

This means only two parameters, the surface tension coefficient σ_{wn} and the static contact angle α_S of the phase with the lower color value are input parameters for this dynamic two–phase model, which is now capable of describing the contact with a third, solid phase. Note that by this definition $\boldsymbol{\nu}_a$ always points away from the phase with the lower color value, tangential to the interface of the solid phase and the phase with the higher color value, which is in the example of Fig. 3.6 the non–wetting phase.
Using the directional derivative the volume reformulation of the contact line force as

described by Eq. (3.40) reads as

$$\delta_{wns_a} = -2\,\hat{\boldsymbol{d}}_a \cdot \sum_b V_b \left(\delta'_{wns_b} - \delta'_{wns_a} \right) \widetilde{\nabla}_a \widetilde{W}_{ab} \qquad (4.14)$$

with

$$\delta'_{wns_b} = \begin{cases} \delta'_{wns_a} & \text{if } b \in fluid \\ 0 & \text{if } b \in boundary \end{cases} \qquad (4.15)$$

and

$$\delta'_{wns_a} = \hat{\boldsymbol{\nu}}_a \cdot \boldsymbol{n}_a \,. \qquad (4.16)$$

Due to the above chosen definitions, δ'_{wns_a} is always positive ($\delta'_{wns_a} > 0$) and the directional derivative towards the solid phase consequently is negative. This fact is in turn compensated by the preceding "-" sign in Eq. (4.14). In this way δ_{wns_a} is positive ($\delta_{wns_a} > 0$) and the volume reformulation for the contact line force is given by

$$\boldsymbol{F}^{vol}_{wns_a} = \boldsymbol{f}_{wns_a} \delta_{wns_a} \,. \qquad (4.17)$$

An illustration thereof is shown in Fig. 4.1(b). The contact line force is zero, if the dynamic (actual) contact angle is equal to the static contact angle and the stronger the deviation to the static contact angle, the stronger the contact line force. The boundary particles can either be described by fixed solid particles or by ghost particles, which are explained now.

4.4 Boundary conditions

The boundary treatment is a challenge in every SPH model. There are different ways how boundary conditions can be implemented. One aspect therefore is the geometry and complexity of the boundary structure. Depending on this given condition along with the corresponding approximation error, the chosen realization of the boundary treatment has a significant influence on the solution, not only close to the boundary, but also within the computational domain, because the implementation may influence the stability of the method, see Basa et al. [Bas09]. Especially if a boundary is not discretized by particles,

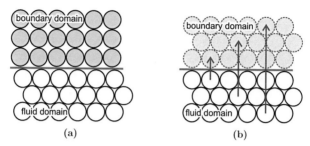

(a) (b)

Figure (4.2): Different boundary methods: (a) Fluid particles move while boundary is discretized by fixed solid particles. (b) Boundary particles dynamically generated by mirroring of fluid particles, known as ghost particle method [Cum99, Sze12a].

the evaluation close to the boundary would be buggy due to the missing complete kernel support by the particle deficiency. Several higher order methods were proposed to improve convergence and accuracy in these regions, cf. Belytschko et al. [Bel98]. The higher the considered order for approximation the better the result, but in turn the higher the computational effort as well. This is why with respect to correction techniques, in this work, the method of Bonet and Lok [Bon99] is chosen, which represents a satisfying compromise of computational effort and increased accuracy.

This method compensates for the particle deficiency close to a boundary and in this manner represents an appropriate way for instance to model a Neumann boundary condition like the normal of an interface close to the contact line.

4.4.1 Velocity boundary condition

There are plenty approaches how boundaries can be realized in SPH. Depending on the case of application a complex boundary treatment may become necessary. Such an example of explicit boundary treatment can be found in Ferrand et al. [Fer13a], Mayrhofer et al. [May13] and Vacondio et al. [Vac12], respectively. In this work two different types are implemented: (a) fixed solid particles and (b) mirrored ghost particles, cf. Fig. 4.2.

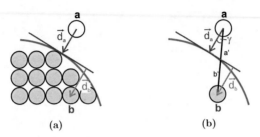

Figure (4.3): No–slip implementation by Morris et al. [Mor97]: Visualization of (a) the principle at a curved fluid–solid interface and (b) the necessary quantities for the scheme of approximation.

Fixed discretized solid particles

This method was published by Morris et al. [Mor97] and is based on a distance weighted counter velocity

$$v_b = -\frac{|d_b|}{|d_a|}v_a \qquad (4.18)$$

to achieve a no–slip condition at the fluid–solid interface. The principal scheme is visualized with the necessary quantities in Fig. 4.3. The boundary particles b are fixed and in this way a fictitious velocity is imposed on them by applying

$$v_{ab} = -\beta v_a\,, \qquad \text{with } \beta = min\left(1.5\,, 1 + \frac{|d_b|}{|d_a|}\right)\,. \qquad (4.19)$$

From Fig. 4.3(b) the implementation is derived. According to the interception theorem with $a' = |d_a|/\cos(\gamma)$ and $\cos(\gamma) = \hat{d}_a \cdot \hat{r}_{ab}$ the ratio $|d_b|/|d_a|$ is obtained by

$$\frac{|d_b|}{|d_a|} = \frac{b'}{a'} = \frac{|r_{ba}| - a'}{a'} = \frac{|r_{ba}|}{a'} - 1 = \frac{|r_{ba}|\cos(\gamma)}{|d_a|} - 1 = \frac{d_a \cdot r_{ba}}{|d_a|^2} - 1\,. \qquad (4.20)$$

Therewith β can be reformulated to

$$\beta = min\left(1.5\,, -\frac{d_a \cdot r_{ab}}{|d_a|^2}\right)\,. \qquad (4.21)$$

This is the method of choice wherever curved fluid–solid interfaces are preexisting in the computational domain.

Mirrored ghost particles

The second boundary method, as illustrated in Fig. 4.2(b), is the method of choice for plane surfaces and domain boundaries as it allows for exact no–slip conditions right on the interface. To impose reference values through a Dirichlet condition on a flat interface, Cummins and Rudman [Cum99] proposed the so–called ghost particles method. There fictitious particles are generated at each time step by exactly mirroring the fluid particles over the interface in this way, that every fluid particle has its counter ghost particle, see Fig. 4.4. By prescribing the velocity of the ghost particles via

$$v_b^g = 2v_{ref} - v_b \,, \tag{4.22}$$

a desired reference value v_{ref} can be imposed right on the interface. A more detailed explanation is given by Szewc et al. [Sze12a]. Neumann boundary conditions can be realized for instance in case of $\frac{dv_x}{dn} = 0$ by setting the ghost velocity to

$$v_{b_x}^g = v_{b_x} \,. \tag{4.23}$$

This principle can be adapted for every desired component. In this way also a component–by–component composition of Dirichlet and Neumann boundary condition can be realized

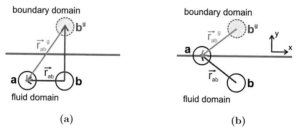

Figure (4.4): Mirroring of ghost particles: (a) Generation of ghost distance vector r_{ab}^g and (b) visualization in the limit of particle a coming close to the boundary itself, illustrating how the two contributions can compensate each other through symmetry.

for vector fields.

Note that using an approach for the momentum balance as published by Adami et al. [Ada10, Ada12], see Eq. (2.90), this formulation for Dirichlet and Neumann boundary conditions is sufficient. The formulation of Szewc et al. [Sze12a], given by Eq. (4.1), however has some peculiarities.

Special treatment of the ghost particle method in combination with the Szewc formulation

The above explained formalism is sufficient for homogeneous flows close to the boundary like the Poiseuille flow but can lead to instabilities when velocity components normal to the wall and irregular particle arrangements are present in the immediate vicinity of the wall in combination with the viscous formulation of Szewc et al. [Sze12a]. An explanation is given in the following. Eq. (4.1) is now reduced to the substantial contribution without kernel and gradient correction for the boundary interaction. The summation is split in a sum over the regular fluid neighbours $\sum_b$ and a sum over the mirrored ghost neighbours $\sum_{b_g}$ by

$$
\begin{aligned}
(\nabla(\eta\nabla \cdot \boldsymbol{v}))_a &= \sum_{b \cup b^g} \xi_{ab}\boldsymbol{v}_{ab} \cdot r_{ab}\nabla_a W_{ab} \\
&= \sum_b \xi_{ab}\left(v_{ab_x}r_{ab_x} + v_{ab_y}r_{ab_y}\right)\nabla_a W_{ab} + \sum_{b^g} \xi_{ab}^g\left(v_{ab_x}^g r_{ab_x}^g + v_{ab_y}^g r_{ab_y}^g\right)\nabla_a W_{ab}^g .
\end{aligned}
$$

$$(4.24)$$

Note that particle a is also reflected at the domain boundary and its ghost particle belongs to the group of neighboring ghost particles b^g.

The ghost distance vector $\boldsymbol{r}_{ab}^g$ of particle a and ghost particle b^g is calculated according to Fig. 4.4(a). With respect to a compelling evidence this formalism is now considered in the limit of particle a sitting right at the interface, cf. Fig. 4.4(b). In this way several simplifications can be applied. The distance vector of the ghost particle can be formulated by

$$
\boldsymbol{r}_{ab}^g = \begin{bmatrix} r_{ab_x} \\ -r_{ab_y} \end{bmatrix} .
$$

$$(4.25)$$

This directly leads to $\xi_{ab}^g = \xi_{ab}$ and Eq. (4.24) becomes

$$
\left. \left(\nabla (\eta \nabla \cdot \boldsymbol{v}) \right) \right|_{wall_a} = \sum_b \frac{\xi_{ab}}{r_{ab}} \left(v_{ab_x} r_{ab_x} + v_{ab_y} r_{ab_y} \right) \begin{bmatrix} r_{ab_x} \frac{\partial W}{\partial r} \\ r_{ab_y} \frac{\partial W}{\partial r} \end{bmatrix}
$$
$$
+ \sum_b \frac{\xi_{ab}}{r_{ab}} \left(v_{ab_x}^g r_{ab_x} - v_{ab_y}^g r_{ab_y} \right) \begin{bmatrix} r_{ab_x} \frac{\partial W}{\partial r} \\ -r_{ab_y} \frac{\partial W}{\partial r} \end{bmatrix} . \tag{4.26}
$$

In the way described above, a no–slip boundary condition would be achieved by setting $\boldsymbol{v}_b^g = -\boldsymbol{v}_b$, which results in

$$
\left. \left(\nabla (\eta \nabla \cdot \boldsymbol{v}) \right) \right|_{wall_a} = \sum_b 2 \frac{\xi_{ab}}{r_{ab}} \begin{bmatrix} \left(v_{a_x} r_{ab_x} - v_{b_y} r_{ab_y} \right) r_{ab_x} \frac{\partial W}{\partial r} \\ \left(v_{a_y} r_{ab_y} - v_{b_x} r_{ab_x} \right) r_{ab_y} \frac{\partial W}{\partial r} \end{bmatrix} . \tag{4.27}
$$

As example, the case, where the velocity in the domain is given by $\boldsymbol{v} = [v(\boldsymbol{r}), 0]$, is now considered. In this way the y–component of Eq. (4.27) results in a in general non–zero value, which means the fluid particles and the ghost particles don't compensate each other in y–direction and therefore the fluid particles can penetrate the boundary. To get rid of this and realize a real no–slip condition, which prevents the particle from penetrating, the boundary condition in terms of the ghost velocity is constituted by $\boldsymbol{v}_b^g = [-v_{b_x}, v_{b_y}]$ with respect to the x–component and set to $\boldsymbol{v}_b^g = [v_{b_x}, -v_{b_y}]$ for the boundary formulation of the y–component. By this

$$
\left. \left(\nabla (\eta \nabla \cdot \boldsymbol{v}) \right) \right|_{wall_a} = \sum_b 2 \frac{\xi_{ab}}{r_{ab}} \begin{bmatrix} v_{a_x} r_{ab_x} r_{ab_x} \frac{\partial W}{\partial r} \\ v_{a_y} r_{ab_y} r_{ab_y} \frac{\partial W}{\partial r} \end{bmatrix} \tag{4.28}
$$

is obtained, which depends only on the constant $\boldsymbol{v}_a$ and has no net y–component for $\boldsymbol{v} = [v(\boldsymbol{r}), 0]$.

Note that this result in square brackets is now comparable to the result of Eq. (2.90) right at the boundary without the modification of the boundary condition.

This mirror approach can be used in general for other quantities like density or pressure as well. The latter is explained in the following.

4.4.2 Pressure boundary condition

The description of pressure in the incompressible SPH scheme is split. First the pressure field is obtained through the pressure Poisson Eq. (2.69). Afterwards in the corrector step the gradient of the pressure field is calculated in the momentum balance to obtain the acceleration due to pressure, see Eq. (2.70). In this way also the boundary conditions have to be implemented in both steps. To incorporate a Dirichlet boundary condition with a reference value of $p = p_{ref}$ in the left–hand side of the numerical pressure Poisson Eq. (4.4), the interaction of ghost particles is added to the regular interaction with

$$p_{ab}^g = p_a - p_b^g = p_a - 2p_{ref} + p_b \,. \tag{4.29}$$

Hence, the pressure Poisson equation reads

$$\sum_b \chi_{ab}(p_a - p_b) + \sum_{bg} \chi_{ab}^g(p_a + p_b) =$$
$$\frac{1}{\Delta t} \sum_b V_b \boldsymbol{v}_{int_{ba}} \cdot \widetilde{\nabla}_a \widetilde{W}_{ab} + \frac{1}{\Delta t} \sum_{bg} V_b \boldsymbol{v}_{int_{ba}}^g \cdot \widetilde{\nabla}_a \widetilde{W}_{ab}^g + \sum_{bg} 2\chi_{ab}^g p_{ref} \,. \tag{4.30}$$

Note the term of the reference pressure p_{ref}, which is shifted to the right–hand side as the whole term represents known quantities. In the same way a Neumann boundary condition with $\frac{dp}{dn} = 0$ is realized through

$$p_{ab}^g = p_a - p_b^g = p_a - p_b \,, \tag{4.31}$$

which leads to the equation

$$\sum_b \chi_{ab}(p_a - p_b) + \sum_{bg} \chi_{ab}^g(p_a - p_b) =$$
$$\frac{1}{\Delta t} \sum_b V_b \boldsymbol{v}_{int_{ba}} \cdot \widetilde{\nabla}_a \widetilde{W}_{ab} + \frac{1}{\Delta t} \sum_{bg} V_b \boldsymbol{v}_{int_{ba}}^g \cdot \widetilde{\nabla}_a \widetilde{W}_{ab}^g \,. \tag{4.32}$$

For both formulations of the left–hand side the right–hand side by Eq. (4.5) requires a boundary condition for the divergence of the velocity field as well. This is realized through

$$\boldsymbol{v}_{int_b}^g = \begin{bmatrix} v_{int_b,x} \\ -v_{int_b,y} \end{bmatrix} \tag{4.33}$$

for a horizontal boundary as shown in Fig. 4.4. Periodic boundary conditions with an optional pressure offset Δp can be applied in the same way by using

$$p_{ab}^p = p_a - p_b^p = p_a - (p_b + \Delta p) = p_a - p_b - \Delta p. \tag{4.34}$$

Applying these conditions gives rise to a linear equation system with the pressure values as the unknown quantity.

For fixed solid particles the pressure can also be solved on these nodes just as if they would be part of the fluid. This way it is also ensured, that no fluid particle will penetrate the wall, because the pressure gradient will keep the fluid particles in the fluid domain, enforced by the incompressibility condition.

After solving the system, the resulting pressure values are used in the corrector step through Eq. (4.2).

4.4.3 Boundary condition for surface tension

To prevent the particles from penetrating the wall due to the continuum surface force, the contribution normal to the wall has to vanish

$$\boldsymbol{d}_a \cdot \boldsymbol{F}_{wn_a}^{vol} \overset{!}{=} 0. \tag{4.35}$$

Morris et al. [Mor97] introduced the interaction of fluid particles with fixed solid particles. This scheme is used for the velocity boundary conditions to ensure $v = 0$ on the interface. Moreover, these boundary particles can also be utilized to generate a weighting function Φ for the implementation of the boundary conditions in terms of the continuum surface force as demanded by Eq. (4.35). The challenge is to introduce a transition in direction of the solid boundary, which yields the same curve for a decreasing continuum surface force as is introduced for an increasing contact line force by the volume reformulation. A possible realization for such a weighting function Φ is given by

$$\Phi_a^i = -\sum_{b^i} V_b \left| \widetilde{\nabla}_a \widetilde{W}_{ab} \right| , \tag{4.36}$$

$$\Phi_a^o = -\sum_{b^o} V_b \left| \widetilde{\nabla}_a \widetilde{W}_{ab} \right| \quad \text{and} \tag{4.37}$$

$$\Phi_a = \left| \frac{\Phi_a^i - \Phi_a^o}{\Phi_a^i + \Phi_a^o} \right| . \tag{4.38}$$

The first equation represents a sum over the interior fluid particles with a weighting by the kernel derivative. In the same way the second equation applies a sum over the outer fixed solid particles. Note that in case of the ghost particle method this formalism can also be applied just by taking the sum over the ghost particles instead of the fixed solid particles. This formulation is chosen in order to be consistent with the second step of the volume reformulation for the contact line force, cf. Eq. (4.14). This way the same shapes for both curves can be achieved. The only difference is the scale with its units. Whereas the quantity for the volume reformulation of the contact line force is predefined by its normalization, the idea behind the concept of Φ_a is to introduce a weighting as a factor to the continuum surface force on a percentage scale. By definition of Eq. (4.38) Φ_a equals 1 if only inner particles are in the vicinity of the considered particle a. The same result would be achieved if the particle would be surrounded just by outer particles, which is irrelevant in this case. But with this formulation a natural transition is generated, which means $\Phi = 1$ as long as particle a is surrounded by fluid particles and as soon as boundary or ghost particles come into play the weighting Φ decreases and finally becomes $\Phi = 0$ as soon as particle a sits right on the interface between a comparable amount of inner and outer particles.

This concept is visualized in Fig. 4.5(a). The origin represents a boundary and the fluid is prescribed on the left side of the boundary. The theoretical interpolated weighting function Φ exactly shows the previously explained behavior from the inner fluid region towards the solid boundary. In order to highlight the interplay with the weighting of the contact line force, Φ is mirrored and put to the same level as δ_{wns}^{1D} by $\Phi \rightarrow -\Phi + 1$. This plot reveals the congruent shape of both weighting functions and in this way confirms this approach for the transition of continuous surface force and contact line force.

Note that usually δ_{wns} is normalized in two dimensions. In a first step δ_{wns}' is achieved by considering the tangential derivative and in the second step δ_{wns} is obained by considering

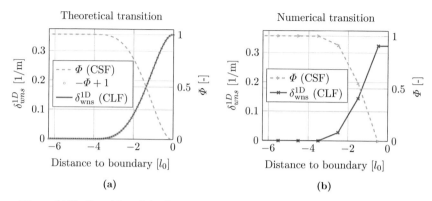

Figure (4.5): Transition of the Continuum Surface Force and the Contact Line Force at a boundary due to their weighting (for illustration in one dimension). The ideal curve of (a) is calculated under the same conditions as the numerical curve of (b) but on continuous space and illustrates the optimum for an adaptive resolution within the original smoothing length.

the derivative in direction normal to the wall. To be able to compare the shape in an one–dimensional illustration, δ'_{wns} is set to $\delta'_{wns} = 1$ here (instead of $\delta'_{wns} = \hat{\boldsymbol{\nu}} \cdot \boldsymbol{n}$). This is just responsible for the scale of δ_{wns} and has no influence on the shape. Therefore, the argumentation is still valid.

In Fig. 4.5(b) the visualization of a real discrete particle distribution is shown. Due to the limiting factor of the particle resolution within the applied smoothing length ($h = 2.1 \cdot l_0$) only three particles are basically used in a radial consideration for the description of the contact line. The only way to overcome this issue would be the introduction of an adaptive spacial resolution. Nevertheless, the results of this approach within this resolution limit are remarkable, as the upcoming chapter will show.

4.5 Time integration by the predictor–corrector scheme

With the discrete SPH formulations of the terms in the momentum balance, the iterative time solution can be constructed. At first, the predictor step is made based on the

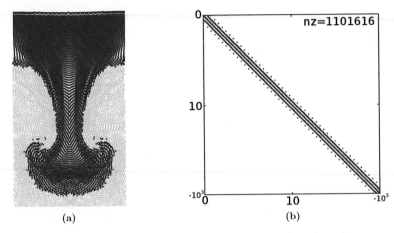

 spans this region; its caption follows.

(a)   (b)

Figure (4.6): Exemplary illustration of non–zero matrix entries for a physical system as shown in (a) with $\approx 20k$ particles and a corresponding number of more than 1 million non–zero matrix elements as shown with blue in (b).

previously mentioned formulation for viscosity, surface tension force and contact line force

$$\boldsymbol{v}_{int_a} = \boldsymbol{v}_{(n-0.5)_a} + \Delta t \left((\nabla(\eta\nabla \cdot \boldsymbol{v})) + \boldsymbol{g} + \frac{1}{\rho}\boldsymbol{F}^{vol}_{wn} + \frac{1}{\rho}\boldsymbol{F}^{vol}_{wns} \right)_a . \tag{4.39}$$

With the intermediate velocity, the pressure Poisson equation has to be solved

$$\frac{1}{\Delta t}\left(\nabla \cdot \boldsymbol{v}_{int}\right)_a = \left(\nabla \cdot \left(\frac{1}{\rho}\nabla p\right)\right)_a . \tag{4.40}$$

An example of such a non–zero matrix structure is visualized in Fig. 4.6. (a) shows the actual state of an exemplary physical system. A two–phase system with a heavier fluid on top of a lighter fluid is illustrated. The heavier fluid is sinking under the influence of gravity. The corresponding matrix non–zero structure is plotted for the same time step in (b). This linear equation system couples the whole computational domain. The utilized libraries are the algebraic multi–grid preconditioner *BoomerAMG* from *hypre* [Fal02, Fal06] and the *BiCGStab* method from *PETSc* [Bal14, Bal13, Bal97].

Note that in the first step of the Leap–Frog formalism Δt becomes $\Delta t/2$ due Eq. (2.74), which has to be adjusted in all equations concerning Δt. With the thereby obtained pressure field, the corrector step is made by

$$\boldsymbol{v}_{(n+0.5)_a} = \boldsymbol{v}_{int_a} - \Delta t \left(\frac{1}{\rho} \nabla p \right)_a . \tag{4.41}$$

Finally, the particle positions are updated through

$$\boldsymbol{r}_{n+1_a} = \boldsymbol{r}_{n_a} + \Delta t \, \boldsymbol{v}_{n+0.5_a} . \tag{4.42}$$

As a consequence a divergence–free velocity field is obtained within the accuracy of the solver in each time step. This is equivalent to an incompressible fluid flow due to the applied constant masses for all particles.

5 Verifications

5.1 Time integration scheme

A reasonable test case for time integration schemes is given by the analysis of periodic systems like a simple pendulum or a spring pendulum. Once elongated and released, the system should always follow the same trajectory with a constant cycle length as long as no friction is present. Moreover, an upper and lower limit of velocity and position in space has to exist, because the total energy can only shift between potential energy and kinetic energy. The new Leap–Frog time integration scheme is now investigated and compared to the well–established Euler method as for instance used in Shao and Lo [Sha03]. In order to make the test case comparable to a spring pendulum, a computational domain with a rectangular shape is used, filled by 100×50 particles, and all particles belong to one phase. On top and bottom a free slip condition is implemented. With respect to pressure, a Neumann boundary condition with $\frac{\partial p}{\partial n} = 0$ at bottom and a Dirichlet condition with $p_{top} = p(x)$ on top are applied. This way the boundary condition on top represents the same pressure gradient as is imposed on the left and right side, see Fig. 5.1. The

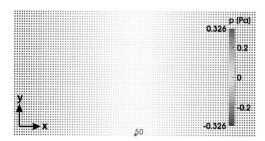

Figure (5.1): Shows the relative pressure field, which varies in time due to the particle elongation with respect to the reference position of the indicator particle.

elongation is measured by the elongation of an indicator particle. Due to the homogeneous motion this could be every particle, here the particle with $ID = 50$ is chosen, cf. Fig. 5.1. The differential equation of a spring pendulum is

$$\frac{\partial^2 x}{\partial t^2} = -\frac{k}{m}(x - x_0) \, .$$

(5.1)

One possible solution therefore is given by

$$x(t) = \hat{x} \sin\left(\sqrt{\frac{k}{m}} t\right) \, .$$

(5.2)

As the sine function repeats every 2π, the cycle time is determined by

$$\sqrt{\frac{k}{m}} T = 2\pi \, .$$

(5.3)

The differential equation of the fluid patch is given by

$$\frac{\partial^2 x}{\partial t^2} = -\frac{1}{\rho} \nabla p \, .$$

(5.4)

In order to make it comparable to the spring pendulum, a linear relationship of the pressure gradient with respect to a particle elongation is introduced by $\nabla p = p_0 \cdot (x - x_0)/L_x$, where L_x represents the length of the computational domain in x–direction. This leads to

$$\frac{\partial^2 x}{\partial t^2} = -\frac{1}{\rho} \frac{p_0 \cdot (x - x_0)}{L_x} \, .$$

(5.5)

By comparing the coefficients of Eq. (5.1) and Eq. (5.5), the cycle time for the "fluid pendulum" can be deduced with

$$T = \frac{2\pi}{\sqrt{\frac{p_0}{\rho L_x}}} \, ,$$

(5.6)

which directly provides the condition for p_0 in dependence of a desired cycle time T. With respect to the conservation of mass, periodic boundary conditions are implemented on left and right side. The total energy for the spring pendulum is given by

$$E_{tot} = \frac{1}{2}mv^2 + \frac{1}{2}k(x - x_0)^2 \,. \tag{5.7}$$

Comparing Eq. (5.3) with Eq. (5.6), a fictitious spring constant k_f can be defined for the fluid pendulum by $k_f = 4\pi^2 \rho V/T^2$. In this manner, the total energy of the fluid pendulum reads

$$E_{tot} = \frac{1}{2}mv^2 + 2\pi^2 \frac{V\rho}{T^2}(x - x_0)^2 \,. \tag{5.8}$$

The result for $T = 1\,\mathrm{s}$, $L_x = 0.01\,\mathrm{m}$ and $\rho = 1000\,\mathrm{kg/m^3}$ is shown in Fig. 5.2. The initial particle x–position is $x = 0.005$ and the origin of the oscillation is set to $x_0 = L_x/3$. Both integration schemes overlap quite well for both position, see Fig. 5.2(a), and velocity, see Fig. 5.2(c), for all shown time integration schemes and applied time step sizes Δt. But with a closer look in the detailed plots of Fig. 5.2(b) and Fig. 5.2(d), the drift in the conventional Euler method is remarkable. The bigger the time step size, the stronger the drift. An even better insight is given by the evolution of the total energy in Fig. 5.3. According to expectations, the Leap–Frog scheme shows an oscillating behavior of the total energy, which is caused by the periodic transfer of kinetic energy to potential energy and vice versa, which is well–known for Verlet methods in periodic regimes. The amplitude is getting smaller with smaller time step sizes. In contrast to the Euler scheme, the total energy of the new Leap–Frog scheme is bounded. A reduction of the time step increment in the Euler scheme leads to a damped increase of the total energy but does not solve the problem with the divergent behavior in general. These findings confirm the numerical implementation, because the illustrated behavior fits the expectations of the first order Euler and the second order Leap–Frog method.

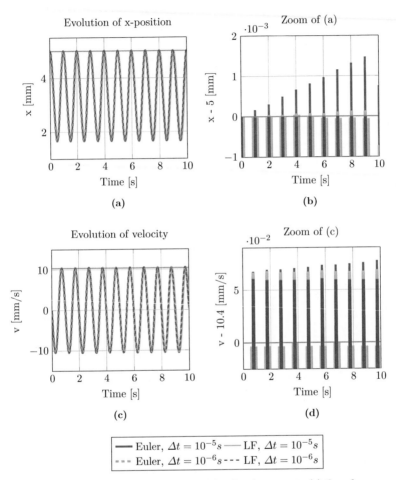

Figure (5.2): Evolution of the "fluid pendulum" with respect to (a) the reference particle position, (b) a detailed plot thereof on the region of interest (x offset shifted by $-5 \cdot 10^{-3}$ m), (c) the velocity of the reference particle and (d) a detailed plot thereof on the region of interest (v offset shifted by $-1.04 \cdot 10^{-2}$ m/s). Results of the Euler scheme and the Leap–Frog scheme are shown for comparison for two different constant time step sizes of $\Delta t = 10^{-5}$ s and $\Delta t = 10^{-6}$ s.

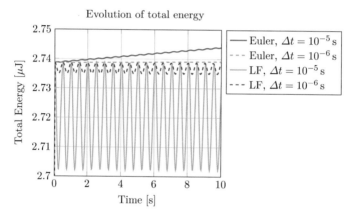

Figure (5.3): Evolution of the total energy for the whole system with two different integration schemes and two different time step sizes each.

5.2 Couette flow

A common test case regarding single–phase or even multi–phase flows is the so–called Couette flow. Here first a single–phase system is setup between two infinite plates, which is realized in the simulation by periodic boundary conditions. The setup is chosen in analogy to the published results by Morris et al. [Mor97]. Therefore the distance between the two plates is set to $L_y = 10^{-3}$ m, kinematic viscosity is set to $\eta = 10^{-6}$ m^2/s and density is set to $\rho = 10^3$ kg/m^3. At time $t = 0$ the system is at rest and the upper plate initially starts to move with a constant velocity of $v_0 = 1.25 \cdot 10^{-5}$ m/s. The computational domain is discretized by 50 particles in every dimension in space. Therewith the Reynolds number becomes

$$Re = \frac{v_0 L_y}{\eta} = 1.25 \cdot 10^{-2}\,. \tag{5.9}$$

The simulation results are compared to the series solution for this case [Mor97]

$$v_x(y,t) = \frac{v_0\,y}{L_y} + \sum_{n=1}^{\infty} \frac{2v_0}{n\pi}(-1)^n \sin\left(\frac{n\pi}{L_y}y\right) \exp\left(-\eta\frac{n^2\pi^2}{L_y^2}t\right)\,. \tag{5.10}$$

In the limit of infinity, this series solution becomes

$$\lim_{t\to\infty} v_x(y,t) = \frac{v_0\,y}{L_y}\,, \tag{5.11}$$

which gives rise to a linear flow profile over the height y. All simulations here are performed with the Leap–Frog time integration scheme, cf. chapter 2.3.2.

The results from Fig. 5.4 reveal, the Szewc et al. [Sze12a] viscosity model reacts more retarded compared to the Adami et al. [Ada10] model, but in the infinite time limit both agree well with the theoretical solution.

To introduce a second fluid layer for the two–phase validation, the same domain is vertically split into two halves. The two fluids are immiscible and the lower fluid 1 and the upper fluid 2 obtain the properties:

$$\eta_1 = 5 \cdot 10^{-7}\,\text{m}^2/\text{s} \qquad \rho_1 = 10^3\,\text{kg/m}^3 \qquad Re_1 = 1.25 \cdot 10^{-2}$$
$$\eta_2 = 1 \cdot 10^{-6}\,\text{m}^2/\text{s} \qquad \rho_2 = 10^3\,\text{kg/m}^3 \qquad Re_2 = 2.5 \cdot 10^{-2}\,.$$

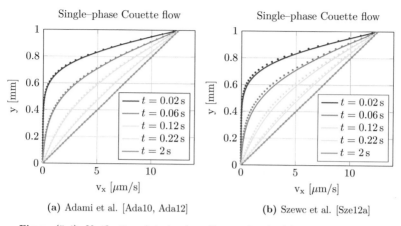

(a) Adami et al. [Ada10, Ada12]

(b) Szewc et al. [Sze12a]

Figure (5.4): Verification of single–phase Couette flow for (a) the viscosity model of Adami et al. [Ada10, Ada12] and (b) the viscosity model of Szewc et al.[Sze12a], cf. chapter 2.4. Markers "●" show the SPH solution whereas straight lines visualize the theoretical solution.

Note that the Reynolds number here is an estimation of the maximal value based on a single–phase flow in the channel to ensure that the conditions for laminar flow are achieved. As both phases will again obtain a linear flow profile, their slope ratio in the velocity profile will be determined by the ratio of their viscosities, which leads to an interface velocity of $v_i = 2/3\,v_0$. The results are shown in Fig. 5.5. Again, in the infinite time limit they both agree with the theoretical solution.

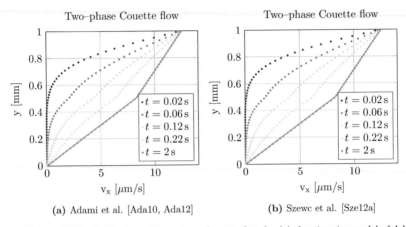

(a) Adami et al. [Ada10, Ada12]   **(b)** Szewc et al. [Sze12a]

Figure (5.5): Verification of two–phase Couette flow for (a) the viscosity model of Adami et al. [Ada10, Ada12] and (b) the viscosity model of Szewc et al.[Sze12a], cf. chapter 2.4. Markers "•" show the SPH solution whereas straight lines visualize the theoretical solution.

5.3 Hagen–Poiseuille flow

As second test case a Hagen–Poiseuille flow profile is validated. In the single–phase flow experiment the same fluid properties and computational domain settings as in the Couette flow were applied, again in analogy to Morris et al. [Mor97]. The system is initially at rest and both plates also stay at rest during the simulation. The fluid is accelerated by a body force through $g_x = 10^{-4}\,\mathrm{m/s^2}$. The theoretical solution for a single–phase Hagen–Poiseuille flow profile is given by [Mor97]

$$v_x(y,t) = \frac{g_x y}{2\eta}(y - L_y) + \sum_{n=0}^{\infty} \frac{4 g_x L_y^2}{\eta \pi^3 (2n+1)^3} \sin\left(\frac{\pi y}{L_y}(2n+1)\right) \exp\left(-\frac{(2n+1)^2 \pi^2 \eta}{L_y^2} t\right). \tag{5.12}$$

In the infinite time limit the solution becomes

$$\lim_{t\to\infty} v_x(y,t) = \frac{g_x y}{2\eta}(y - L_y). \tag{5.13}$$

This leads to a Reynolds number of $Re = 1.25 \cdot 10^{-2}$. All simulations here are again performed with the Leap–Frog time integration scheme, cf. chapter 2.3.2.

Fig. 5.6 illustrates how both models perform. They are both slightly retarded compared to the theoretical solution, but the model of Szewc et al. is first even more retarded and in the infinite time limit this model overshoots the theoretical velocity profile.

In the two–phase experiment the domain is split in the same way as described for the two–phase Couette flow and the two immiscible fluids have the same properties:

$$\eta_1 = 5 \cdot 10^{-7}\,\mathrm{m^2/s} \qquad \rho_1 = 10^3\,\mathrm{kg/m^3} \qquad Re_1 = 1.25 \cdot 10^{-2}$$
$$\eta_2 = 1 \cdot 10^{-6}\,\mathrm{m^2/s} \qquad \rho_2 = 10^3\,\mathrm{kg/m^3} \qquad Re_2 = 2.5 \cdot 10^{-2}\,.$$

Note that again the Reynolds number is an estimation based on Eq. (5.13), which provides the maximal velocity, to ensure the fluid flow is laminar.

The fluids are accelerated by $g_x = 10^{-4}\,\mathrm{m/s^2}$. The overhoot of the Szewc et al. model in the infinite time limit is also visible in the two–phase results as shown in Fig. 5.7. Additionally, it is observed in accordance with theory that the vertical position of the peak

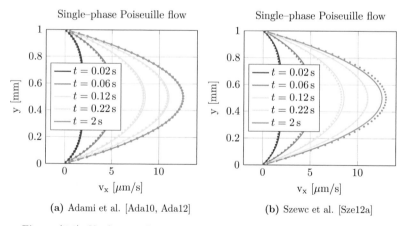

(a) Adami et al. [Ada10, Ada12]   (b) Szewc et al. [Sze12a]

Figure (5.6): Verification of single–phase Hagen–Poiseuille flow for (a) the viscosity model of Adami et al. [Ada10, Ada12] and (b) the viscosity model of Szewc et al.[Sze12a]. Markers "•" show the SPH solution whereas straight lines visualize the theoretical solution.

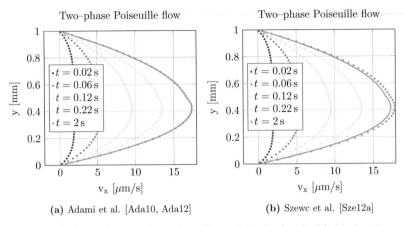

(a) Adami et al. [Ada10, Ada12]   **(b)** Szewc et al. [Sze12a]

Figure (5.7): Verification of two–phase Hagen–Poiseuille flow for (a) the viscosity model of Adami et al. [Ada10, Ada12] and (b) the viscosity model of Szewc et al.[Sze12a]. Markers "•" show the SPH solution whereas straight lines visualize the theoretical solution.

velocity is shifted into the lower, less viscid phase. The theoretical results for comparison of the steady state solution are calculated according to Reis and Phillips [Rei07] by

$$v_{x_1} = \frac{g_x h_c^2}{2\eta_1} \left(-\left(\frac{y'}{h_c}\right)^2 + \frac{y'}{h_c}\left(\frac{\mu_1 - \mu_2}{\mu_1 + \mu_2}\right) + \frac{2\mu_1}{\mu_1 + \mu_2} \right) \quad \forall \quad -h_c \leq y' \leq 0, \quad (5.14)$$

$$v_{x_2} = \frac{g_x h_c^2}{2\eta_2} \left(-\left(\frac{y'}{h_c}\right)^2 + \frac{y'}{h_c}\left(\frac{\mu_1 - \mu_2}{\mu_1 + \mu_2}\right) + \frac{2\mu_2}{\mu_1 + \mu_2} \right) \quad \forall \quad 0 \leq y' \leq h_c, \quad (5.15)$$

where h_c represents the half channel width $h_c = L_y/2$ and y' is the new control variable, which has its point of origin in the middle of the channel.

In general, the results agree quite well with the theoretical model, especially the formulation of Adami et al. [Ada10] in case of the single–phase flow, see Fig. 5.6(a), and the two–phase flow, see Fig. 5.7(a).

The complete studies with respect to the ghost particle method were performed for the Euler scheme as well. The results can be looked up in the "Semesterarbeit" of V. Hägele [Häg15]. Validations for fixed solid particles were performed in the master thesis of

T. Woog [Woo11]. As they don't show any essential information on top of the presented results, they are not incorporated in this work.

5.4 Rayleigh–Taylor instability

A further common test case regarding immiscible two–phase flow is the Rayleigh–Taylor instability. To setup an experiment a confined domain by $0 < x < 1\,\mathrm{m}$ and $-1 < y < 1\,\mathrm{m}$ with no–slip conditions at the domain boundaries was chosen in analogy to Cummins and Rudman [Cum99], Hu and Adams [Hu07] and Szewc et al. [Sze12b]. The system was setup for a Reynolds number of [Sze12a]

$$Re = \frac{\sqrt{L^3 g}}{\eta} = 420\,. \tag{5.16}$$

The initial interface is set by $y_I(x) = 0.15 \sin(2\pi x)$ and is visualized in Fig. 5.8(a) and Fig. 5.8(e). The properties for the lower fluid 1 and upper fluid 2 are set to:

$$\eta_1 = 7.45 \cdot 10^{-3}\,\mathrm{m^2/s} \qquad \rho_1 = 100\,\mathrm{kg/m^3}$$
$$\eta_2 = 7.45 \cdot 10^{-3}\,\mathrm{m^2/s} \qquad \rho_2 = 180\,\mathrm{kg/m^3}\,.$$

Through the gravitational acceleration the initial perturbation, which is imposed on the interface by the sine function, is in the shear field further enhanced. In this way the instability rises. The evolution of this system is on the one hand modeled with SPH, see (a)–(d), and on the other hand modeled with OpenFOAM®, see (e)–(h). In general, the results are again in good agreement, which verifies the implementation of the SPH model. With respect to (d) and (h) the methodical difference is revealed, as the higher level of detail in the Volume of Fluid (VOF) results is visible compared to the SPH results. This is in general attributed to the smoothing of the SPH approach. To obtain the same details in SPH, the resolution has to be increased, but thereby the computational effort would increase as well.

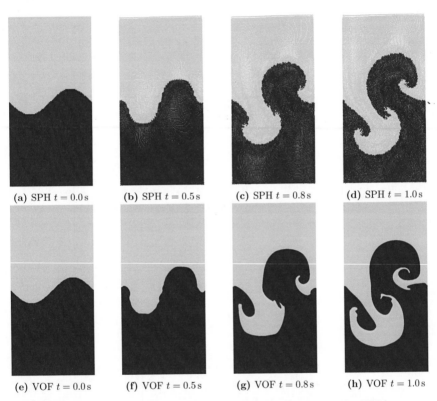

Figure (5.8): Rayleigh–Taylor instabilities with no–slip walls calculated by SPH (upper row, 20k particles) and VOF (lower row, 32k cells). VOF plots are performed with Open-FOAM®. The heavier fluid (ρ_1) is based on top and the density ratio is set to $\rho_1/\rho_2 = 1.8$. Gravitational acceleration is applied by $g = 9.81\,\text{N/kg}$ and Reynolds number is set to $Re = 420$.

5.5 Surface tension

In the next step, the test case for the model of immiscible two–phase flow is extended in order to describe interface and contact line phenomena arising from surface tension. First, the focus solely is on interfaces and investigations are made for a system at rest and under dynamic conditions. Afterwards, the influence is analyzed with respect to contact lines, which yields static contact angles for systems at rest and dynamic contact angles for systems under motion.

5.5.1 Interfaces

Static evaluation

As first verification an isotropic drop is considered at rest, see Fig. 5.9(a). There are no other external influences on the drop except that surface tension is present. The equilibrium state is achieved by relaxing the system for 0.25 s. In this way the particle arrangement overcomes the initial underlying cartesian grid structure and the particles at the interface arrange in a more homogeneous round shape. After relaxation the radius of the drop is determined to $R = 0.187\,\text{m}$ by the discrete step of the color function. Due to the Young–Laplace equation for an isotropic sphere, cf. Eq. (3.11), the pressure jump at the interface can be calculated and serves as static validation with respect to theory. Fig. 5.9(b) shows the comparison of the Continuum Surface Force model and the Continuum Surface Stress formulation along with the theory. Due to the smoothing of SPH a transition zone is introduced around the physical interface as expected. This representation is achieved by projecting the values of each node in cartesian description to the radial component of a polar coordinate system. Thus, the origin of the polar coordinates is set to the middle of the drop and the theoretical relative pressure jump is calculated by $\Delta p = \sigma / R$. This way the pressure inside the drop is given by $p = 5.35\,\text{Pa}$ and the reference pressure outside the drop is here set to $p = 0$ by definition.

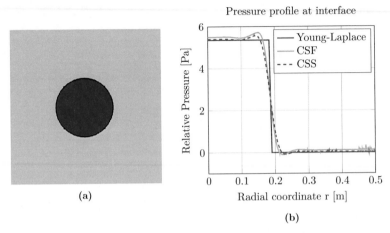

(a)

(b)

Figure (5.9): Verification of the surface tension model for a system at rest. (a) shows an isotropic two–dimensional drop and (b) the corresponding pressure jump at the interface for different models. The domain properties are set in analogy to [Mor00, Ada10]. Hence the size is given by $1x1$ m with a discretization of $100x100$ particles and for simplicity $\sigma = 1$ is assumed.

Dynamic evaluation

After this comparison for the system at rest, a dynamical test case is performed now. The same system is considered, but after the first 250 ms a velocity field is imposed on the drop by

$$v_x = v_0 \frac{r_x}{r_0} \left(1 - \frac{r_y^2}{r_0 r} \right) \exp\left(-\frac{r}{r_0} \right), \tag{5.17}$$

$$v_y = v_0 \frac{r_y}{r_0} \left(1 - \frac{r_x^2}{r_0 r} \right) \exp\left(-\frac{r}{r_0} \right), \tag{5.18}$$

with $v_0 = 10$ m/s, $r_0 = 0.05$ m and $r = \sqrt{r_x^2 + r_y^2}$. Fig. 5.10(a) shows the prescribed flow field. This is again performed in accordance to [Mor00, Ada10] but with a two–phase system confined by no–slip walls on top and bottom. Due to the symmetry of the system, the periodic boundary conditions on left and right side also ensure that particles don't cross

Evolution of drop radius

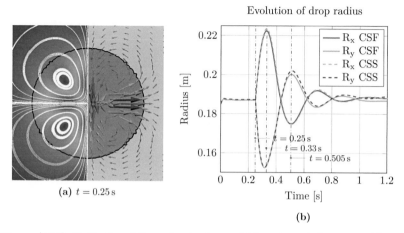

(a) $t = 0.25\,\text{s}$

(b)

Figure (5.10): Verification of the surface tension model for a dynamical test case. (a) shows the imposed velocity field. In the left half streamlines visualize the flow field whereas velocity vectors are used for illustration in the right half. The color index represents the relative strength of the velocity field. (b) shows the temporal evolution of the different drop radii.

the boundaries. Both densities are set to $\rho_1 = \rho_2 = 1\,\text{kg/m}^3$ and the dynamic viscosities are prescribed by $\mu_1 = \mu_2 = 0.05\,\text{kg/s m}$. The temporal evolution of this system is shown in Fig. 5.10(b). The first $250\,\text{ms}$ are reserved for the relaxation of the system as already addressed in the use case of the static validation. Then with the elongation by the simultaneously imposed velocity field, the drop starts to oscillate. For an appropriate analysis the radius is decomposed into a x and y component, which depicts the interface position of the drop with respect to the considered spatial direction. Due to the imposed flow field the component in x–direction becomes bigger and in the same way the y–direction is getting smaller until the first maximum and minimum are reached respectively. Fig. 5.11(a) shows the state of the drop and the corresponding velocity field close to this turning point at time $t = 0.33\,\text{s}$. Accordingly Fig. 5.11(b) shows the state with a maximal elongation in y–direction and minimal extension in x–direction at time $t = 0.505\,\text{s}$. Note that the color index and the velocity vectors are rescaled to visualize the flow field every time step. In an absolute comparison the magnitude of the velocity of course reaches a minimum at

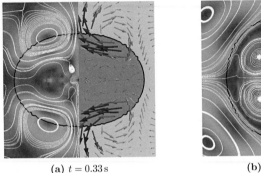

(a) $t = 0.33\,\mathrm{s}$ (b) $t = 0.505\,\mathrm{s}$

Figure (5.11): Visualization of the drop state very close to the turning point. (a) shows the elongation with a maximum radius in x–direction and (b) the elongation with a maximum radius in y–direction.

turning points of these radii.

The previous examples verify the implementation with respect to interface dynamics. The results confirm the expectations as these models are already known for several years. Based on this, now the use case is extended in terms of a third solid phase, which gives rise to distinct wetting phenomena accompanied by distinct contact angles.

5.5.2 Contact angles

Static evaluation

Before dynamic wetting conditions are investigated, the model is again first examined with respect to static equilibrium states. Hence, the relaxation of droplets on a solid wall is considered first, cf. Fig. 5.12. The simulation starts with an initial configuration as shown in (a). For a better focus solely on the wetting model both phases have the same densities and viscosities. As soon as the system is released, surface tension is driving the system towards an energetic optimum, which means the predefined static contact angle is achieved associated with a minimization of the two–phase interfacial area. Consequently the interface obtains a homogeneous curvature in such a way that no net force is left and the systems ends up in its equilibrium state.

In this new approach the contact line force is responsible for the motion of the contact

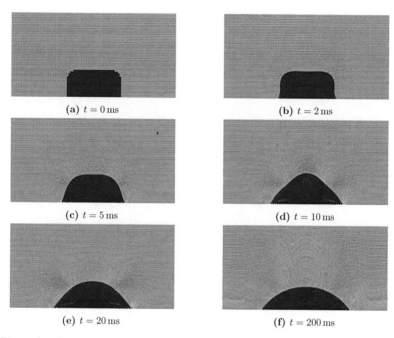

Figure (5.12): Relaxation of a two–phase system on a wall. Both phases have the same densities $\rho_1 = \rho_2 = 1000\,\text{kg/m}^3$, the dynamic viscosities are set to $\mu_1 = \mu_2 = 0.01\,\text{kg/(s\,m)}$ and the surface tension is set to $\sigma_{wn} = 0.0182\,\text{N/m}$ with a static contact angle of $\alpha_S = 50°$. (a) Shows the initial state, and further states are shown after (b) 2 ms, (c) 5 ms, (d) 10 ms, (e) 20 ms and finally (f) the system at rest after 200 ms. The computational domain is set to $L_x = 0.01\,\text{m}$, $L_y = 0.005\,\text{m}$ and is discretized by 200x100 particles.

line and in this way adjusts the contact angle. The magnitude of this force is depending on the difference of actual and static contact angle. In this way an initial non–equilibrium contact angle gives rise to a contact line force, which causes the motion of the contact line until the equilibrium contact angle is reached. The transition from the initial contact angle to the static contact angle of $\alpha_S = 50°$ is made by a motion of the contact line, which causes dynamic contact angles. This temporal evolution is shown in Fig. 5.12(b)–(e). The final equilibrium state with the droplet at rest is shown in (f). Fig. 5.13 shows the results of other simulations for a range of static contact angles reaching from $30° - 140°$. Their

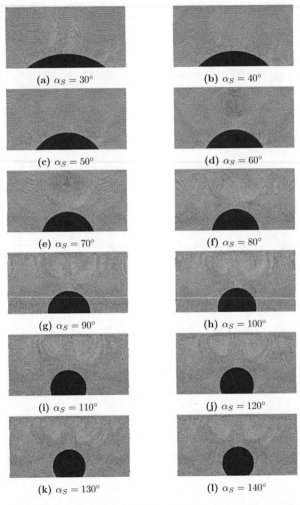

(a) $\alpha_S = 30°$ **(b)** $\alpha_S = 40°$

(c) $\alpha_S = 50°$ **(d)** $\alpha_S = 60°$

(e) $\alpha_S = 70°$ **(f)** $\alpha_S = 80°$

(g) $\alpha_S = 90°$ **(h)** $\alpha_S = 100°$

(i) $\alpha_S = 110°$ **(j)** $\alpha_S = 120°$

(k) $\alpha_S = 130°$ **(l)** $\alpha_S = 140°$

Figure (5.13): Equilibrium drop shapes (state after $1\,\mathrm{s}$), evaluated with $20k$ particles. Both phases have the same densities $\rho_1 = \rho_2 = 1000\,\mathrm{kg/m^3}$ and the dynamic viscosities are set to $\mu_1 = \mu_2 = 0.01\,\mathrm{kg/(s\,m)}$.

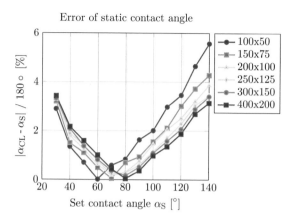

Figure (5.14): Error of the measured static contact angle α_{CL} with respect to the predefined contact angle α_S using the ghost particle method for different particle resolutions.

absolute equilibrium states errors compared to the predefined contact angles are shown in Fig. 5.14. The error in this partial wetting regime is below 5% and it is observed that the shape of the error plot is getting more and more symmetric, the higher the resolution is. The comparison of different initial grid sizes in this figure reveals that the error of the static contact angle does not decrease but converge while increasing the particle resolution. This is because the detection of the contact angle and the unit normal at the contact line is independent on the particle resolution as the same absolute amount of particles within one cut–off radius is responsible for the contact angle of all resolutions. The error is given by the relative particle arrangement and in this way given by the geometry of the interface. If the particle resolution is changed, the particle arrangement at a given contact angle stays the same as long as no adaptive spacial resolution is introduced with more particles belonging to one cut–off radius at a contact line. It is important to note that through a defined particle arrangement with a limited amount of particles close to the tip of a contact line, only certain configurations can be realized which represent certain contact angles and that there is for this reason a trade–off in achieving a constant density field and resolving a thin contact angle. In addition, the contact line is losing its "sharpness" the flatter the interface is, because in this way, the area, which obtains a interface normal, is broadened. Within this broader area the values for interface normal and contact angle stronger vary, which is an additional reason for the increasing error for strong wetting or non–wetting regimes in Fig. 5.14. A rough estimate on the amount of particles, which represent a contact line, can be made by using half of a 2D kernel support as indication of the volume for a contact line, cf. Fig. 3.3(d). As the kernel here is applied with a smoothing length of $h = 2.1l_0$, where l_0 is the initial particle spacing, the amount of particles in the volume of a kernel is $\pi \, (2 \cdot 2.1)^2 \approx 55$. This means a contact line is due to the missing complete kernel support roughly described by 27 particles, but approximately only half of them in turn have a significant influence based on the volume reformulation of the contact line force. In this way only few particles contribute to the contact line and increasing the resolution admittedly has an influence on the computational domain, but has only indirectly an influence on the description of the boundary and the contact line. Due to the smooth contact line area in the simulation, a single contact angle is determined through

$$\alpha_{CL} = \frac{1}{\sum_b \delta_{wns_b}} \sum_b \alpha_b \, \delta_{wns_b} \, . \tag{5.19}$$

In this way the information of particles in the vicinity of a contact line is merged into a single value representing the contact angle of the physical interface.

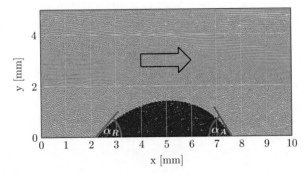

Figure (5.15): Exemplary steady state droplet shape under constant motion to the right by a pressure gradient of $\Delta p = -4 \, \text{Pa}$ over $\Delta x = 0.01 \, \text{m}$.

Dynamic evaluation

In the next step, contact angle and contact line are analyzed under dynamic conditions. After the drop has reached its equilibrium state, a pressure gradient at $t = 0.25$ s is applied, to show steady state advancing and receding contact angles under a constant motion of the drop, see Fig. 5.15 as an excessive example with a gradient of $\frac{dp}{dx} = -400$ Pa/m. The temporal evolution of the advancing and receding contact angle for a static contact angle of $\alpha_S = 50°$ by a pressure gradient of $\frac{dp}{dx} = -200$ Pa/m along with the corresponding apex height is shown in Fig. 5.16(a). It is observed that the left and the right contact angle of the drop obtain the same values, and as soon as the pressure gradient is switched on, the lines for advancing and receding contact angles separate. The evolution of the corresponding contact line velocity is shown in Fig. 5.16(b). The contact line velocity is hereby determined through

$$r_{CL} = \frac{1}{\sum_b \delta_{wns_b}} \sum_b r_b \, \delta_{wns_b} \quad \text{and} \quad v_{CL} = \frac{dr_{CL}}{dt}. \tag{5.20}$$

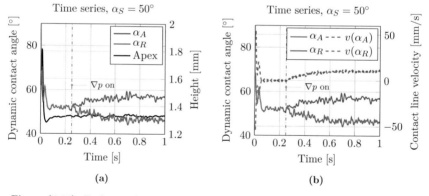

(a)   (b)

Figure (5.16): Evolution of advancing and receding contact angle in time for a static contact angle of $\alpha_S = 50°$ and $\nabla p = -200$ Pa/m. In (a) additionally the evolution of the drop apex is shown and in (b) the corresponding advancing and receding contact line velocities are plotted.

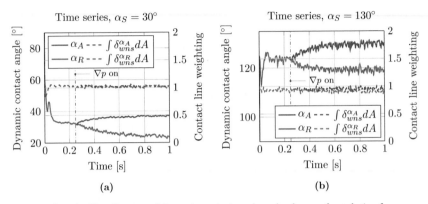

Figure (5.17): Visualization of dynamic contact angle and volume reformulation for contact lines δ_{wns} as defined by Eq. (3.40). Evolution for (a) $\alpha_S = 30°$ and (b) $\alpha_S = 130°$ showing the conservation of the integral condition of Eq. (3.37) while the contact angle itself changes the value.

In this example both contact lines move simultaneously so that the drop does not further spread on the solid wall after the equilibrium state is left. As verification of the volume reformulation for contact lines δ_{wns}, the evolution of $\int \delta_{wns}\, dV$ is plotted through the whole process in Fig. 5.17, where (a) serves as validation for a strong wetting regime and (b) validates the non–wetting regime. Both figures reveal that the volume reformulation fulfills its demands as written in the integral condition of Eq. (3.37). The integral value is preserved while the contact angle itself changes its value. The low resolution of the contact line by only few particles is responsible for the jittering behavior, but nevertheless the model proves to be quite reliable. This simulation, as shown in Fig. 5.16 and Fig. 5.17, is now performed for a complete range of static contact angles from $\alpha_S = 30° - 140°$, each one of it for a set of pressure gradients with $\frac{dp}{dx} = $ -500; -400; -300; -200; -100; 0 Pa/m. In this manner each simulation provides a steady state droplet shape with one advancing and one receding contact angle. Fig. 5.18(a) illustrates the dynamic behavior for a static contact angle of $\alpha_S = 50°$ and reveals the model dependent linear relationship of the dynamic contact angle on the contact line velocity in this approved range around the static contact angle. The results for all steady state contact angles are plotted in Fig. 5.19. In this way each marker of (a) together with one of (b) correspond to one simulation.

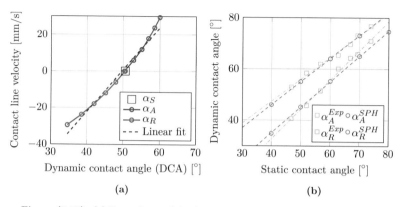

Figure (5.18): (a) Dependence of the dynamic contact angle on the contact line velocity for a static contact angle of $\alpha_S = 50°$. (b) Advancing and receding contact angles in dependency of the static contact angle. Experimental values taken from Lam et al. [Lam02]. SPH simulation carried out with $\sigma_{wn} = 0.0182\,\mathrm{N/m}$ and $\nabla p = -200\,\mathrm{Pa/m}$. Dashed lines represent linear fits of the corresponding experimental/simulated values.

An experimental validation is made with the results published by Lam et al. [Lam02]. In order to obtain dynamic contact angles, they placed a droplet on a solid substrate and changed the volume through a needle, which is introduced from the bottom of the solid substrate. The height of the tip is adjusted to end up right above the solid–liquid interface in the middle of the base area of the droplet. Dynamic contact angles are then measured by varying the volume of the droplet and looking at the contact angle of the moving contact line. While increasing the volume the advancing contact angle is measured and while decreasing the volume the receding contact angle is measured. The results of both experiment and simulation are plotted with good agreement in Fig. 5.18(b). Note that due to the linear dependency of the contact angle on the contact line velocity the simulation results could easily be shifted, but it confirms that for a comparable system setup advancing and receding contact angle are similar in experiment and simulation. Moreover the graph reveals the right trend with respect to different wetting conditions.

It is also possible to perform dynamic simulations with predefined static contact angles below 30° and above 140°. The contact line force will calculate an appropriate driving force, but the contact line will be badly discretized. Additionally, due to the narrow tips of

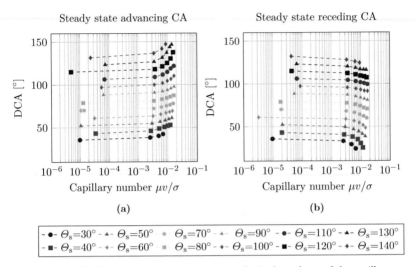

Figure (5.19): Steady state dynamic contact angles in dependence of the capillary number, (a) the advancing contact angle and (b) the receding contact angle.

the contact angles, the region of the contact line will become very broad in strong wetting or non–wetting regimes. In this broader area the interface normal and consequently the contact angle will vary stronger. The closer the dynamic contact angle reaches a value of $\alpha_D = 0°$ or $\alpha_D = 180°$ the less the contact angle determinable. But as mentioned above, a driving force with these values would be possible as long as the system does not really reach the equilibrium state.

Last but not least a "grid convergence" with simulations of different initial particle resolutions for the same physical setup is shown, cf. Fig. 5.20. For this purpose a static contact angle of $\alpha_S = 50°$ is again chosen and the droplet is accelerated by a pressure gradient of $\nabla p = -200 \, \mathrm{Pa/m}$ just as in the previous examples. Note that, according to Fatehi et al. [Fat11], a grid convergence with an irregular particle distribution may fail as long as no higher order methods are applied. A demonstration thereof is also published by Tartakovsky et al. [Tar15]. The graph reveals the convergence for both advancing and receding contact angle and confirms along with the previous validations the successful implementation of this new model.

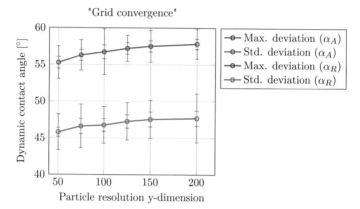

Figure (5.20): "Grid convergence" for different particle resolutions and a static contact angle of 50°. Error bars show the maximum deviation and standard deviation with respect to the time averaged mean value.

6 Applications

In the previous chapters the model was built and verified for a range of test cases. Static interface shapes with related pressure jumps were compared to theory and the droplet dynamics with respect to an elongation by an imposed flow field was analyzed. After the verification of static and dynamic contact angles, the applicability of the model is demonstrated now.

6.1 Qualitative wettability simulations on curved solids

An interesting application is the wettability of porous structures. In a two–phase system there's always a wetting phase and a non–wetting phase, which means the void space can be differently occupied depending on the wetting conditions of the involved fluids along with the applied process conditions. In usual drainage experiments the void space in porous structures is initially saturated with the wetting fluid and then displaced under pressure by a non–wetting fluid, cf. the work of S. Dwenger [Dwe12, Dwe15] and Kunz et al. [Kun15]. Afterwards, this process can be reversed, which is then called imbibition. The output of these experiments gives some indication about the macroscopic characteristics of the whole system including the porous structure, the surface constitution and the utilized fluids. A complete description of these physical relations is not intended in this work. Detailed physical models in this context can be looked up in literature, e.g. de Gennes et al. [deG03] and in Pinder and Gray [Pin08]. Instead, in this section a proof of concept shall be shown to illustrate the applicability of the introduced two–phase model in the context of fluid dynamics in porous solids.

6.1.1 Determining equilibrium states in porous materials

There are different ways to determine equilibrium states of two–phase systems in porous structures [Dwe15]. For instance in the simulated annealing method, the energy of the system is minimized based on a Monte Carlo approach by swapping single discretized fluid elements or clusters. In principle, these swaps are stochastically initiated, but only accepted if an energetic optimization can be achieved. This gives rise to a two–phase distribution, which may be optimal from an energetical point of view, but this distribution may also be never reached during the course of a real wetting and dewetting process respectively. The benefit of the simulation method presented in this work is, that this model is capable of describing wetting states depending on real process conditions.

To demonstrate the general usability and show qualitative wettability simulations on curved solids, a use case as shown in Fig. 6.1 is realized. The initial state is visualized in (a). The properties of both fluids are set to $\rho_1 = \rho_2 = 1000 \, \mathrm{kg/m^3}$, $\mu_1 = \mu_2 = 0.1 \, \mathrm{kg/s\,m}$ and the surface tension is prescribed by $\sigma_{wn} = 0.07 \, \mathrm{N/m}$. An external gravitiational field is applied by $g = 9.81 \, \mathrm{m/s^2}$ and the domain size with $L_x = 14.1 \, \mathrm{mm}$ and $L_y = 13.2 \, \mathrm{mm}$ is discretized by $141x132$ particles, which implies an initial particle spacing of $l_0 = 0.1 \, \mathrm{mm}$. Within the first 200 ms the prescribed contact angles are set and the equilibrium state is visible. Fig. 6.1(b) shows the equilibrium state in case of a blue, wetting fluid with a static contact angle of $\alpha_S = 40°$. Due to the different sphere radii and the resulting different pore throats between two spheres, the curvatures of the fluid–fluid interfaces vary. In this way stronger curvatures are obtained for smaller throats. As both densities are the same, there's no counter force like a hydrostatic pressure gradient in terms of a capillary rise experiment. Consequently surface tension is not only driving the motion but is also responsible for the position of the interface at rest. The geometric setup is symmetric with regard to a verticle axis in the middle of the plot, which means the equilibrium state is this state, where the curvature of the upper and the lower interface are the same. Naively spoken this would be the case, when the interfaces have reached a height, where both pore throat diameters are the same. Actually, it's a little more complex. Due to a rise of the contact line on the solid sphere also the orientation of the solid tangent changes by the round shape of the sphere. This additionally causes the curvature of the upper interface to decrease. In this way an equilibrium state is already achieved before both interfaces reach a height, where the geometrical diameters of the upper and lower pore throat are the same.

Fig. 6.1(c) reveals the equilibrium state in case of a blue non–wetting fluid with a static contact angle of $\alpha_S = 140°$. The same argumentation is valid with respect to final interface positions. Not only the geometrical diameter of the pore throat between two spheres is responsible for the equilibrium interface position, but also the orientation of the surface

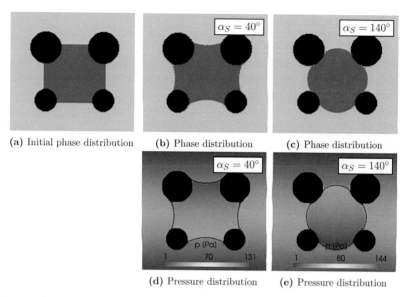

(a) Initial phase distribution **(b)** Phase distribution **(c)** Phase distribution

(d) Pressure distribution **(e)** Pressure distribution

Figure (6.1): Determination of equilibrium states for two–phase interfaces enclosed by solid spheres. (a) shows the initial state, (b) the equilibrium state for the blue wetting fluid and (c) the equilibrium state in case of a non–wetting fluid. (d) and (e) visualize the corresponding pressure distribution for the wetting and non–wetting blue fluid in equilibrium.

tangent on the solid sphere. Initially, the upper and lower interfaces with their contact lines were oriented right in the vertical middle of the spheres. But then, as the upper throat is thinner, the enclosed blue non–wetting bubble is slightly pushed towards the bottom until the previously explained equilibrium state is reached.

Figs. 6.1(d) and (e) show the corresponding relative pressure states, where the offset is given by the reference pressure on top of the domain. The hydrostatic pressure field is visible by the linear pressure increase outside the pore. Both plots illustrate how the pressure behaves due to surface tension and in which way pressure jumps are obtained due to the curvature of the interfaces.

6.1.2 Rising bubbles in porous structures

In the next step a densitiy ratio of $1/4$ is applied in order to obtain a general motion through a solid structure. Therefore, first a bubble is placed on bottom of the computational domain, see Fig. 6.2. For comparison three different wetting conditions are simulated and their time series are shown in: Figs. (a)–(d) for a wetting bubble with a static contact angle of $\alpha_S = 47°$, Figs. (e)–(h) for a less wetting bubble with a static contact angle of $\alpha_S = 74°$ and Figs. (i)–(l) for a non–wetting bubble with a static contact angle of $\alpha_S = 133°$. The bubble density is set to $\rho_1 = 500 \,\mathrm{kg/m^3}$, whereas the other fluid obtains a density of $\rho_2 = 2000 \,\mathrm{kg/m^3}$. Both dynamic viscosities are set to $\mu_1 = \mu_2 = 0.01 \,\mathrm{kg/s\,m}$ and the surface tension is prescribed by $\sigma_{wn} = 0.03 \,\mathrm{N/m}$. The prescribed wetting condition is valid for the solid spheres and the bottom as well. Through the external gravitiational field with $g = 9.81 \,\mathrm{m/s^2}$ the lighter bubble is accelerated towards the top of the computational domain, which is dimensioned by $L_x = 20 \,\mathrm{mm}$, $L_y = 30 \,\mathrm{mm}$ and discretized by $200 x 300$ particles.

The different wetting scenarios of Figs. (a), (e) and (i) already show on bottom a different remaining covered surface area of the bubble. The stronger the wetting condition the bigger the covered surface area at time $t = 50 \,\mathrm{ms}$. Further it can be deduced that bubbles with less wetting affinity detach earlier. Another significant difference with respect to the detachment over the whole time is given by the fact, that for the non–wetting case with $\alpha_S = 133°$ no residue of the lighter fluid remains on bottom. This means the stronger the wetting the more residue remains on bottom. In the next column at time $t = 100 \,\mathrm{ms}$ a further characteristic behavior of wettability becomes visible. The wetting bubbles in Figs. 6.2(b) and (f) are literally attracted by the solid spheres, whereas the non–wetting bubble in (j) is avoiding the contact with the solid structure. This leads to a general difference in the way how the lighter fluid is moving through the solid structure. The upcoming plots of the time series reveal that in the wetting case the bubble is following a way along the solids and in this way is more or less building bridges to overcome the gaps between the solids, in contrast to the non–wetting case. Here the bubble transport takes place in a more classical sense just like a flow in a channel with no–slip walls.

In general, the simulation results show in this context that with respect to mass transport in porous media, stronger wetting fluids tend to leave more residue on the solid structure compared to less wetting fluids. Moreover, the wetting condition may have a strong influence on the propagation of the bubble and in this way on the flow field during the whole process.

A real experiment of multi–phase mass transport in porous media including a pseudo three–dimensional description is published in Kunz et al. [Kun15]. The drainage experiments in this work are on the one hand simulated with the surface tension model presented here and on the other hand measured in the laboratory with well–defined micro–models. The geometrical interface positions are in a good agreement with respect to experiment and simulation. But a major discrepancy is revealed in terms of the dynamical, temporal evolution of the system. In principle two reasons are invoked to explain this difference. The first one is about the experimental setup, which may exert a higher flow resistance due to connections in the fluid conduit. The second, and even more important in this context, is the so–called stick–slip effect [Bla02, Ren11], which is originating from surface impurities and heterogeneities. They are not taken into account in this model but may have a severe influence on the temporal evolution of the system and moreover, are responsible for a hysteretic contact angle as defined in chapter 3.2.1.

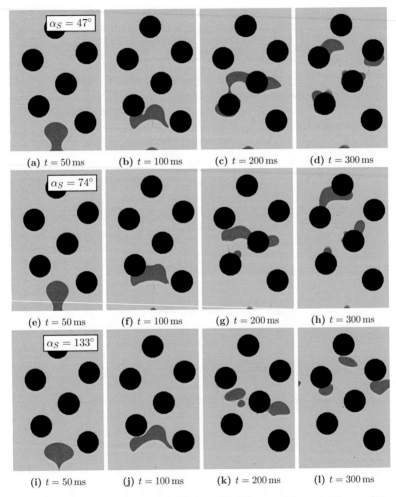

Figure (6.2): Rising bubbles in a liquid column for different arbitrary wetting conditions. Figs. (a)–(d) show the time series for a blue wetting fluid with $\alpha_S = 47°$. Figs. (e)–(h) reveal the temporal evolution for $\alpha_S = 74°$ and Figs. (i)–(l) shed light on the process for a non–wetting fluid with $\alpha_S = 133°$.

6.2 Primary bubble formation at an orifice

This model for surface tension addresses in principle a wide range of possible applications. Therefore, beside the previously presented flow in porous media, other areas are conceivable. For applications with dominating surface tension forces it's usually sufficient to implement a model with static contact angles. The advantage of this new surface tension model is, it is not only able to describe applications with static contact angles, but furthermore able to describe the full range of flow regimes up to the viscous flow dominated regime, where surface tension has no influence anymore. In this way it can also handle the transition area from a surface tension dominated to a viscous flow dominated regime. In this area dynamic contact angles occur. One example with this purpose is a bubble formation process at an orifice of bubble column reactors. Depending on the process condition, the prerequisites for dynamic contact angles are met, leading to different bubble dynamics and bubble volumes compared to models with static contact angles.

In many industrial applications with regard to gas–liquid or liquid–liquid reactions, bubble and droplet formation respectively play an important role. They may be initially formed right at the inlet of sparged stirred tank reactors, spray reactors or column reactors [Hen13] and the initial bubble or droplet size is one essential quantity, which influences the overall process and effects product quality and reaction yield.

6.2.1 Background

In the past there were many investigations made concerning gas–liquid reactions in bubble column reactors. Note that the references here shall show the research areas in association with the challenges of bubble column reactors and do not claim for completeness.

Some of the first attempts were made in the late '50s, where Govier et al. [Gov57, Gov58] studied the general behavior of upwards vertical two–phase flow in terms of air–water mixtures. The experiments revealed, that the flow pattern is anything but simple due to the influence through slip, holdup and the general unsteady process conditions. Since these days, empirical correlations for predictive descriptions [Sha82, Wil92] are subject of investigations. In a further step, theoretical one–dimensional models were built, which qualitatively showed an agreement of experiment and simulation [Fle96]. Due to the increasing computational power, later, two–dimensional models became subject of new investigations. First Euler–Euler methods were considered, where each phase is described

as a continuous phase [Sok94, Sok04]. Through the general unsteady flow profile time–
dependent transient simulations became inevitable [Bec94]. Many of the nowadays applied
numerical models for multi–phase flow are based on a Volume of Fluid approach (VOF)
[Sca99, Try11], which describes the interface of two phases by a smooth transition function
with a finite thickness. This method became a well–engineered method for the simulation
of three–dimensional bubbly flows [Ann05]. Therewith the influence of the flow pattern
on reactive transport of gas–liquid systems may be examined [Bot04]. One interesting
effect of different flow patterns are distinct bubble wake types, which depend on process
conditions, e.g. the surrounding flow field. This is why neighboring bubbles may also have
an influence on each other and bubble swarms therefore give rise to unique characteristic
wake types as well [Koy06]. As most of the gas–liquid mixing is happening in bubble
wakes during the rise, these different wake types have a significant influence on reaction
yield and selectivity [Koy04]. In this way expensive direct numerical models can be used
to build and feed reduced models like an effective film model, which describes the thin
volume around a single bubble, cf. [Rad08].

Beside the development of physico–chemical models, the reduction of the computational
effort is a further scientific challenge. In Euler–Lagrange approaches the disperse fluid
phase is described by a Lagrangian particle, which is not further spatially resolved as it
would be the case in an Euler–Euler consideration. Therefore effective coupling conditions
have to be introduced but with the overall benefit of gaining computation time. With
respect to bubble columns, this is an appropriate model reduction, because the mobility of
the gaseous phase is high, which would lead more or less to an instantaneous equalization
of an inhomogeneous concentration profile within a gaseous bubble. Beyond that, the
reaction takes place in the wake of the bubble in the liquid phase. Thus, there's no need
to discretize the bubble as a continuum and an effective film model can be utilized to
effectively take the reactive transport into account [Rad10].

Yet another enhancement with respect to computational efficiency can be achieved by
using an appropriate turbulence model like a $k - \epsilon$ model [Pfl01] or a Large Eddy Simulation (LES) [Dee01]. Moreover, a Detached Eddy Simulation (DES) can save additional
computation time [Mas14] while still preserving the demanded physical accuracy in the
domain of interest.

In terms of numerical discretization schemes progress has also been made and alternative
models to the Finite Volume Method (FVM) and VOF are still under investigation. If
complex boundary structures and especially complex moving interfaces are considered,

grid–based methods become quite challenging due to the necessary remeshing and a possible adaptive mesh refinement, which dramatically increases computation time in transient simulations. Examples for such a use case are coalescence and breakup of bubbles, respectively. Particle–based, mesh–free methods like Smoothed Particle Hydrodynamics (SPH) [Mon05] may be beneficial when dealing with these issues. The bubble detachment from a submerged orifice using the continuum surface force is modeled with SPH in [Das09]. Later, their model was improved by taking contact angles into account with the diffuse interface (DI) method [Das10, Das11]. The process of rising bubbles modeled by SPH was investigated in detail by Grenier et al. [Gre13]. Moreover, a comparison for coalescing bubbles with the Level–Set method is shown in this work.

An overview of state of the art techniques for bubbly flows with bubble formation, coalescence and breakup is published in [Yan07]. Two main gas flow regimes are distinguished when a gaseous phase is injected from a single orifice: the bubbling regime and the jetting regime. As surface tension phenomena are investigated the focus in this work is on the bubbling regime, which is obtained at lower gas flow rates. Single bubble formation is also examined in Ma et al. [Ma12]. They used a two–dimensional VOF method with a model for a static contact angle and analyzed the influence of surface tension, density, viscosity and inflow velocity. A comparable three-dimensional setup in combination with experimental validations are shown in Yujie et al. [Yuj12]. Their analysis with respect to surfaces tension just focuses on the interaction at the two–phase interface, but in general surface tension also plays an important role at the three–phase contact line area. It affects the dynamic wetting or non–wetting behavior, resulting in various dynamic contact angles, which in turn are responsible for the initial bubble formation as will be shown later.

There are models with constitutive equations for dynamic contact angles, see Sikalo et al. [Sik05]. With the herein introduced surface tension model however, dynamic contact angles are obtained as a result of the momentum balance. Therefore no constitutive equation and no further process parameters are necessary for the simulation. In the application here for bubble columns flow regimes at the orifice above a capillary number of $Ca > 10^{-4}$ are considered, which induces the usage of a dynamic contact angle, as the apparent contact angle deviates from the static contact angle, cf. Fig 5.19. Through the changed boundary condition by the dynamic contact angle, curvature and pressure drop at the two–phase interface consequently change as well, leading to a different bubble formation process compared to a model with static contact angles.

As previously described, bubble volume and bubble shape play an important role with regard to conversion and reaction yield. The bubble is formed at the orifice and fed by the injected gas. In this way the initial bubble extension on the orifice and the bubble volume are implicitly predefined by fluid and process properties and moreover by the geometry of the orifice and its wetting properties. The dimensioning of a bubble column reactor therefore has to start at the dimensioning of the capillary and the orifice. With a well–designed orifice a primary bubble size distribution can be adjusted in the bubble column reactor, which is in turn the basis for effective models with regard to yield and conversion. In this manner the primary bubble volume has to be the first controlled quantity in the overall process. As long as a critical Weber number of $We < 2$ is not exceeded, single bubble formation takes place, see Mersmann et al. [Mer77]. Räbiger and Vogelpohl [Rae81] showed by recordings of a high speed camera, that beyond a Weber number of $We > 2$ the primary bubbles are split into two halves due to the snap–off at the orifice. This leads to a reduction of the bubble diameter by $\approx 21\%$. Klug and Vogelpohl [Klu83] also investigated the secondary bubble formation and derived a physical relation for the secondary bubble diameter. As Blaß [Bla88] reported, a further effect on the process of secondary bubble formation arises from multi–injection sites through the complex shear field caused by other bubbles. By further increasing the inflow volume flux, coalescence and breakup may happen in the jetting regime. Beyond that the Plateau–Rayleigh jet instability influences the bubble formation due to the breakup of the jet. For even higher velocities atomization may occur, see [Bla88].

In the bubbling regime however, where no coalescence and breakup happens, the initial bubble volume is an important quantity to characterize the overall process. Bubble velocity and gas holdup respectively can be deduced from the size of the primary bubble, see Shah et al. [Sha82]. The gas holdup is a measure for the gas volume fraction in the bubble column. In combination with the mean bubble volume and the mean detachment frequency the mean residence time is given. An accumulated interfacial area can be calculated from the mean bubble diameter and the gas holdup. Together with the mean residence time and appropriate reaction kinetics, the conversion for the overall process can be calculated. This is why the process of single bubble formation and the detachment dynamics at the inlet are interesting for the overall process. In the above described low flow regime, surface tension and buoyancy are the competitive and dominating forces.

The new contribution in this work is, that orifice geometries like capillary, nozzle and diffuser are investigated under the influence of their wetting conditions. This gives rise to different bubble volumes, which are neglected by today's effective models. The findings of this chapter are partly published in Huber et al. [Hub16a].

6.2.2 Problem setup

The geometry of the computational domain is designed to focus on primary bubble formation. As already mentioned, the flow field has an influence on the bubble detachment. Therefore the idea is to keep the flow field in the vicinity of the inlet clean from external influences or complex flow conditions, which would cause a flow feedback on the bubble formation process and therefore give rise to varied bubble sizes and dynamics.

The setup along with the applied boundary conditions is shown in Fig. 6.3(a). To limit the computational domain size while still realizing a representative wider physical domain, periodic boundaries are applied on left and right side. A Dirichlet boundary condition

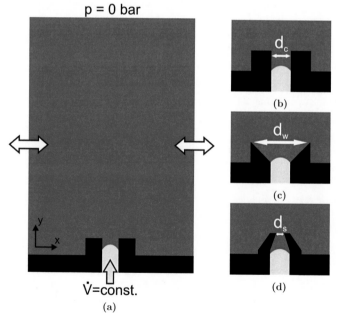

Figure (6.3): (a) Computational domain with boundary conditions and different orifice geometries: (b) a regular capillary, (c) a diffuser and (d) a real jet nozzle.

for the pressure is incorporated at the top, which consequently means a divergence–free velocity field has to be imposed on top. Otherwise, the incompressibility condition in terms of the pressure Poisson equation would be over–determined. For that reason a Neumann boundary condition for velocity is applied. The boundary conditions at the top of the computational domain reads:

$$p\big|_{top} = 0 \tag{6.1}$$

$$\frac{\partial v_y}{dy}\bigg|_{top} = 0 \tag{6.2}$$

$$\frac{\partial v_x}{dy}\bigg|_{top} = 0\,. \tag{6.3}$$

At the inlet a constant volume flow rate for the incoming gaseous phase shall be achieved. Usually this would mean to define a matching inflow velocity. But the velocity profile in such a capillary is due to the Hagen–Poiseuille flow profile not constant across the main flow direction. Moreover, due to feedback of the flow field from the bubble column during the course of the simulation, a better way to achieve a certain volume flux is to apply a Dirichlet pressure boundary condition at the bottom and to introduce a PI–controller, which utilizes the pressure gradient of inlet and outlet to impose the demanded volume flux at the inlet. As a consequence, the complete velocity field inside the capillary is still a solution of the momentum balance and is not influenced by a Dirichlet boundary condition. In this way the pressure at the inlet becomes a function of the actual flow rate $\dot{V}_{act}$ and the desired flow rate $\dot{V}_{set}$. Therewith the boundary conditions at the narrow cross–section of the inlet is given by:

$$p\big|_{in} = p\left(\dot{V}_{act}, \dot{V}_{set}\right) \tag{6.4}$$

$$\frac{\partial v_y}{dy}\bigg|_{in} = 0 \tag{6.5}$$

$$\frac{\partial v_x}{dy}\bigg|_{in} = 0\,. \tag{6.6}$$

In a Lagrangian description, where the discretization is done in mass elements and not in space, an inflow and outflow condition implies to introduce new discretization nodes in terms of particles at the inlet and removing them, when they are about to leave the domain. These boundary conditions for the so called open boundaries from Kunz et al.

$L_x\,[mm]$	$L_y\,[mm]$	$d_c\,[mm]$	$d_w\,[mm]$	$d_s\,[mm]$	$\#x$	$\#y$	$l_0\,[mm]$
20.0	40.0	3.0	6.0	1.4	200	400	0.1

Table (6.1): Dimensions of the computational domain and different orifice diameters.

[Kun16] have a special feature. They are capable of handling inflow and outflow at the same time in the same plane depending on the flow field. And this behavior can change dynamically based on the course of simulation. This capability is especially necessary at the top of the domain, where outflow happens through the rising gas bubbles and inflow may occur due to the induced backflow of the flow field in the surrounding area of the bubble. The physical extension of the computational domain including the dimensions of the different inlet geometries are given in Tab. 6.1. The domain size is initially resolved by 200×400 particles in x and y direction, which corresponds to an inter–particle spacing of $l_0 = 1e^{-4}\,\mathrm{m}$.

Numerical model

To be capable of modeling high density ratios, the viscous model of Adami et al. [Ada10] is utilized as basic formulation of the momentum balance. This means the pressure gradient of the momentum balance is formulated by Eq. (2.89) and the viscosity term is discretized by Eq. (2.90). To model density jumps at the interface, the multi–phase formulation for the density field by Eq. (2.56) is applied and all derivatives are built using the first–order consistent formulation of Eq. (2.39). The discrete formulation of the pressure-Poisson Eq. (2.69) in this model reads

$$\frac{2}{V_a} \sum_b \frac{(V_a^2 + V_b^2)\,p_{ab}}{\rho_a + \rho_b} \frac{\boldsymbol{r}_{ab} \cdot \widetilde{\nabla}_a \widetilde{W}_{ab}}{|\boldsymbol{r}_{ab}|^2 + \zeta^2} = \frac{1}{\Delta t} \sum_b V_b \left(\boldsymbol{v}_{int_b} - \boldsymbol{v}_{int_a} \right) \cdot \widetilde{\nabla}_a \widetilde{W}_{ab}\,. \tag{6.7}$$

The surface tension model is implemented as described in chapter 4.3. Due to the high density ratios in this application a further stabilization of the interface is introduced. In a surface tension model, the pressure gradient and the surface tension forces play a key role. In equilibrium, they have to cancel each other. In order to realize this in a smooth numerical description right at an interface, it is necessary to obtain comparable accelerations of both fields but oppositely directed, which means both smooth force distributions should be subject of the same profile. The pressure field is obtained in the pressure Poisson equation, cf. Eq. (6.7), by a combination of a smooth SPH derivative and a finite difference approximation. Afterwards, the acceleration is calculated by a further smooth formulation of the pressure gradient. Consequently a corresponding formulation for the color function has to be found, because the color function defines the distribution of the surface tension force, cf. Eqs. (3.20), (3.21) and Fig. 3.3(d). This means at an interface the pressure field and the transition of the color function should be subject of the same slope. Therefore a suitable modification for the smooth derivative of the color function has to be developed. In this way the original formulation, cf. Eq. (4.6), is replaced by a new description for the interface normal

$$ \boldsymbol{n} = \frac{1}{V_a[c]} \sum_b \bar{c}_{ab} \left(V_a^2 + V_b^2 \right) \widetilde{\nabla} \widetilde{W}_{ab} . \tag{6.8}$$

with an effective color value $\bar{c}_{ab}$, defined by

$$ \bar{c}_{ab} = \frac{\bar{\rho}_b c_a + \bar{\rho}_a c_b}{\bar{\rho}_b + \bar{\rho}_a} . \tag{6.9}$$

In contrast to the original multi–phase description for a weakly compressible SPH scheme by Adami et al. [Ada10], the gradient is slightly reduced through

$$ \bar{\rho}_a = \rho_a - \frac{\rho_a - \rho_b}{\omega} , \quad \bar{\rho}_b = \rho_b \quad \forall \quad \rho_a > \rho_b . \tag{6.10}$$

In this way the gradient of the color function better fits the smoother pressure gradient of the incompressible SPH scheme. ω is a fitting quantity, which can be adjusted to obtain a stable interface. The limits are $\omega = 1$, which corresponds to a non–density weighted gradient, and $\omega \to \infty$, which equals the Adami approach. In this work a value of $\omega = 1.2$ empirically prove to be appropriate. Further explanations and visualizations of the modified color function can be looked up in the Bachelor's thesis of D. Dobesch [Dob15].

This modification together with the model and findings of the previous chapters are implemented in the SPH Code SiPER [Dob15, Kun15, Hir16], which is commonly developed at the Institute of Chemical Process Engineering at the University of Stuttgart and used for further calculations.

6.2.3 Bubble detachment results

With the above described model and boundary conditions, the influence of different flow rates and wetting conditions on primary bubble formation is now investigated for the distinct orifice types shown in Fig.6.3(b)–(d).

Propagation of the two–phase system

In the examined simulations a density ratio of $1/100$ is applied, which e.g. corresponds to an air–water system under a pressure of $10\,\text{bar}$ and the viscosity ratio is set to $1/10$. The surface tension coefficient of the gas–liquid interface is set to $\sigma = 0.055\,\text{N/m}$. In this way various gas–liquid and liquid–liquid systems can be realized just by adapting the necessary fluid properties. The volume flux at the inlet is defined by the average inflow velocity of $0.075\,\text{m/s}$. The complete fluid properties are listed in Tab. 6.2. A time series under these conditions for the capillary inlet is shown Fig. 6.4. (a)–(c) show the evolution

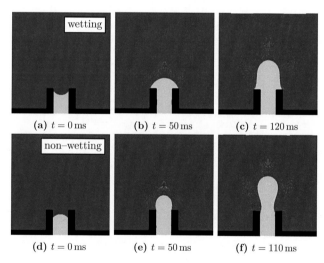

Figure (6.4): Time series with capillary inlet of the first bubble formation with an inlet velocity of $0.075\,\text{m/s}$. (a),(b) and (c) correspond to the wetting case with a contact angle of $45°$ for the injected fluid. (d),(e) and (f) correspond to a non–wetting case with a contact angle of $135°$.

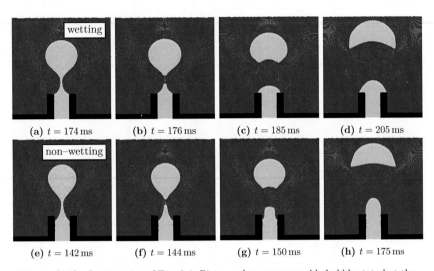

(a) $t = 174\,\mathrm{ms}$ **(b)** $t = 176\,\mathrm{ms}$ **(c)** $t = 185\,\mathrm{ms}$ **(d)** $t = 205\,\mathrm{ms}$

(e) $t = 142\,\mathrm{ms}$ **(f)** $t = 144\,\mathrm{ms}$ **(g)** $t = 150\,\mathrm{ms}$ **(h)** $t = 175\,\mathrm{ms}$

Figure (6.5): Continuation of Fig. 6.4. Pictures show a comparable bubble state but the time code reveals a different dynamical propagation. In the non–wetting case (e)–(h) the bubble detaches faster as in the wetting case (a)–(d). Therefore the bubble in the wetting case just becomes bigger, as inflow velocities are the same.

	fluid 1	fluid 2
density ρ_i	$10\,\mathrm{kg/m^3}$	$1000\,\mathrm{kg/m^3}$
dynamic viscosity μ_i	$0.001\,\mathrm{Pa\,s}$	$0.01\,\mathrm{Pa\,s}$
surface tension coefficient σ	$0.055\,\mathrm{N/m}$	

Table (6.2): Fluid properties of injected, dispersed fluid 1 and continuous fluid 2.

in case of an injected wetting fluid with a contact angle of $\alpha_S = 45°$. (d)–(f) show the evolution for a injected non–wetting fluid with a contact angle of $\alpha_S = 135°$. (b) and (e) are made for comparison at the same time step. In the first one the natural wetting behavior is revealed. The injected phase covers the solid surface according to expectations in contrast to the non–wetting case, where the injected fluid only is in contact with the solid phase within the capillary. As a consequence, (c) and (f) already give a hint how the bubble growth differs in both cases. Due to the stronger wetting regime, the extension of the attached gaseous phase on the orifice becomes bigger. The continuation thereof in

Fig. 6.5 further reveals that in the non–wetting regime the snap–off happens earlier, cf. (b) and (f). Knowing that the volume fluxes for both examples are the same, it can be deduced, that with a later snap–off the bubble in the wetting case therefore has the bigger volume. (c) and (d) again confirm the previously described wetting behavior, whereas (g) and (h) show the non–wetting behavior. Actually Figs. 6.4(d)–(f) and Figs. 6.5(e)–(h) even reveal, how the interface in terms of the contact line is pulled back into the capillary in the moment of the snap–off. In the wetting regime the injected fluid level always stays above the capillary ending.

Note that in a given substance system these different wetting behaviors can still be utilized by changing the material of the orifice or by introducing different coatings in order to affect the bubble size. In this way a tuning parameter is given through the wetting conditions even for a fixed orifice geometry.

In Fig. 6.6 the simulation is repeated under the same conditions but with a nozzle inlet.

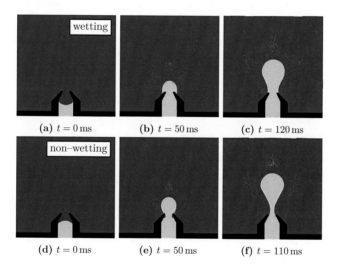

Figure (6.6): Time series of the first bubble formation with nozzle inlet and an inlet velocity of $0.075 \, \text{m/s}$. (a),(b) and (c) correspond to the wetting case with a contact angle of $45°$ for the injected fluid, (d),(e) and (f) correspond to a non–wetting case with a contact angle of $135°$.

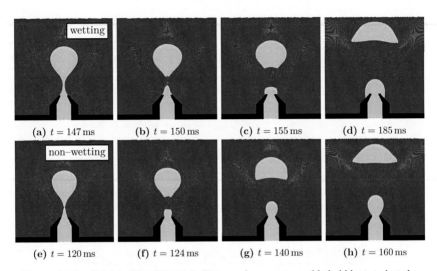

(a) $t = 147\,\mathrm{ms}$ **(b)** $t = 150\,\mathrm{ms}$ **(c)** $t = 155\,\mathrm{ms}$ **(d)** $t = 185\,\mathrm{ms}$

(e) $t = 120\,\mathrm{ms}$ **(f)** $t = 124\,\mathrm{ms}$ **(g)** $t = 140\,\mathrm{ms}$ **(h)** $t = 160\,\mathrm{ms}$

Figure (6.7): Continuation of Fig. 6.6. Pictures show a comparable bubble state but the time code reveals a different dynamical propagation. In the non–wetting case (e)–(h) the bubble detaches faster as in the wetting case (a)–(d). Therefore the bubble in the wetting case becomes bigger, as inflow velocities are the same.

The nozzle has the same cross section at the inlet but is constricted at its outlet. The dimensions are visualized in Fig 6.3 and the corresponding sizes are written in Tab. 6.1. Due to the constriction the velocity, especially for incompressible fluids, has to be higher at the outlet of the nozzle. In the hereby obtained time series Fig. 6.6(e) shows a little initial enclosure of the wetting fluid, which initially wetted the interior wall of the nozzle. Based on the angle of constriction, the wetting conditions and the flow field conditions, this enclosure can be affected. Also the startup process itself has an influence on the formation and continuance of this enclosure in the inlet. The continuation of this simulation is shown in Fig. 6.7. If the wetting behavior is compared to the non–wetting behavior, the simulation for the nozzle basically shows the same trend as the capillary in the considered flow regime. An injected wetting fluid detaches later, which in this way leads to bigger bubbles. A comparison of capillary inlet and nozzle inlet reveals, that bubbles formed by the latter detach earlier for both wetting conditions. In this way the nozzle in general leads to smaller bubble volumes. (a)–(d) also reveal, that due to the low inflow velocity, surface tension is still dominating, because in this illustrated wetting regime, wetting of the outer nozzle happens faster and is therefore stronger than inertia and buoyancy are driving the detachment process.

Last but not least the simulation is again repeated under the same conditions with a diffuser inlet. The results of the first time steps are shown in Fig. 6.8. Again, the cross–section of the diffuser right at the domain boundary has the same extension as the capillary and the nozzle, but then the diffuser becomes wider. The continuation is shown in Fig. 6.9. For the diffuser, again, the wetting case leads to bigger bubbles and by comparing the results to the previous simulations of nozzle and capillary the diffuser leads to the latest detachment times in conjunction with the biggest bubble. In this manner, the order by increasing detachment times reads nozzle, capillary and diffuser. The same ordering applies for increasing bubble sizes. A complete comparison of a range of wetting

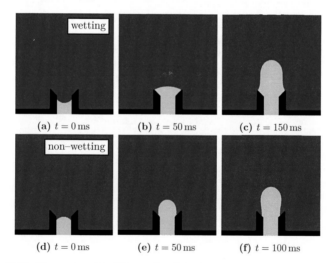

(a) $t = 0\,\mathrm{ms}$ **(b)** $t = 50\,\mathrm{ms}$ **(c)** $t = 150\,\mathrm{ms}$

(d) $t = 0\,\mathrm{ms}$ **(e)** $t = 50\,\mathrm{ms}$ **(f)** $t = 100\,\mathrm{ms}$

Figure (6.8): Time series with diffuser inlet of the first bubble formation with an inlet velocity of $0.075\,\mathrm{m/s}$. (a),(b) and (c) correspond to the wetting case with a contact angle of $45°$ for the injected fluid, (d),(e) and (f) correspond to a non–wetting case with a contact angle of $135°$.

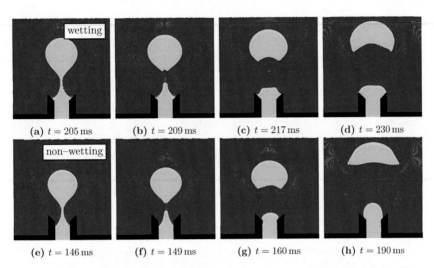

Figure (6.9): Continuation of Fig. 6.8. Pictures show a comparable bubble state but the time code reveals a different dynamical propagation. In the non–wetting case (e)–(h) the bubble detaches faster as in the wetting case (a)–(d). Therefore the bubble in the wetting case just becomes bigger, as inflow velocities are the same.

properties for the three different geometries with different inflow velocities will be drawn now.

Analysis of the bubble detachment

The simulations, as shown in the previous section, are now made for a range of static contact angles from $\alpha_S = 45°$ to $\alpha_S = 135°$ with a step size of $\Delta\alpha_S = 15°$. Additionally, the inflow velocity is varied for each simulation from $v_{in} = 0.035\,\text{m/s}$ to $v_{in} = 0.075\,\text{m/s}$ in steps of $\Delta v_{in} = 0.01\,\text{m/s}$. The results with respect to the primary bubble diameter are shown in Fig. 6.10. The contact angles are given with respect to the injected fluid, denoted as "fluid 1" in Tab. 6.2. Therefore a contact angle of $\alpha_S = 135°$ means a non–wetting, disperse fluid 1 and as a consequence a continuous wetting fluid 2 with a contact angle of $45°$. The influence of wetting and inflow velocity on the capillary inlet is shown in Fig. 6.10(a). In the considered flow regime the results show in general an increase of the bubble diameter associated with an increase of the inflow velocity. According to expectations the diameter can also be increased by increasing the wetting strength. Hence, the experiment with a contact angle of 45° and an inflow velocity of 0.075 m/s leads to the biggest bubbles. For comparison with literature the correlations by Jamialahmadi et al. [Jam01] are plotted in the same graph. They are built by an empirical fit function in terms of

$$d_b = \left(\frac{5}{Bo^{1.08}} + \frac{9.261 Fr^{0.36}}{Ga^{0.39}} + 2.147 Fr^{0.51} \right)^{\frac{1}{3}} \cdot d_c \tag{6.11}$$

to make predictive statements about the bubble size in gas–liquid systems. The function depends on the Bond number Bo (given with respect to the orifice diameter), Froude number Fr and Galilei number Ga, given by

$$Bo = \frac{\rho_l g d_c^2}{\sigma}\,, \tag{6.12}$$

$$Fr = \frac{v_{in}^2}{d_c g} \quad \text{and} \tag{6.13}$$

$$Ga = \frac{\rho_l^2 d_c^3 g}{\mu_l^2}\,. \tag{6.14}$$

The index l labels the liquid phase. This model was fitted to experiments, where wettability and orifice types were not distinguished. A further reference is given by the model of Gaddis and Vogelpohl [Gad86]. They derived a physical approach based on a force balance on the bubble. The approximate solution for this approach reads

$$d_b = \left(\left(\frac{6 d_c \sigma}{\rho_l g} \right)^{\frac{4}{3}} + \left(\frac{81 \mu_l \dot{V}}{\rho_l \pi g} \right) + \left(\frac{135 \dot{V}^2}{4 \pi^2 g} \right)^{\frac{4}{5}} \right)^{\frac{1}{4}}\,, \tag{6.15}$$

which is valid for low and moderate gas pressures with $\dot{V} = \pi d_c^2 v_{in}/4$.

In the considered velocity range this approach leads to almost equal bubble sizes as the implicit formulation of Voit et al. [Voi87], which is also published in the VDI Wärmeatlas [Ver06].

It is observed that the simulated bubble diameters are in the same order of magnitude as the models reported in literature. The new model reveals the same trend with increasing inflow velocity and is moreover capable of describing the influence of wettability and different orifice geometries.

The results for the diffuser are shown in Fig. 6.10(b). These simulations are performed under the same conditions as (a). The plot quantitatively shows, that the diffuser in general yields bigger volumes while the overall trend with respect to the inflow velocity is the same. Correlations from the literature usually don't distinguish between different orifice types and are mainly made for capillary–like needles.

The result for the nozzle are plotted in (c). The overall trend with respect to an increase in the inflow velocity again agrees with literature. Due to the constriction of the nozzle, the fluid velocity in case of the nozzle is higher than in case of the capillary. This may be the reason, why in the diagram the spreading of the bubble diameters caused by the different wetting conditions is less compared to the capillary. Surface tension is losing influence over inertia the faster the inflow velocity is. In turn, this may also be the reason, why in case of the diffuser, due to the lower velocity, the spreading of the bubble diameter is bigger compared to the capillary. As the models from literature are based on a capillary with a single diameter, d_c is replaced by d_s for the correlation of the nozzle. Note that using the correlation with d_c would fit the simulation results quite well.

An overview of the resulting bubble sizes for these three orifice types under different wetting conditions is shown in Fig. 6.11. All experiments here are again made under the same conditions for density ratio, viscosity and volume flux. In (a) the inlet diameter of the orifice type is kept constant with $d_c = 3$ mm. Therefore the inflow velocity right at the inlet of the orifice is due to the same cross–section the same in all three orifice types, giving rise to a common inflow velocity at the inlet of 0.055 m/s. In (b) the outlet diameter of capillary and nozzle are adjusted to the same value of $d_c = 1.4$ mm. Therefore in this case both average outlet velocities are due to the same cross-section at the outlet of capillary and nozzle the same and equal 0.253 m/s. These two plots show again that simulation results and both correlations, originating from a variety of experiments and the reduced physical model, are in the same order with regard to the bubble diameter. But moreover they also reveal, that the orifice type definitely has an influence, as bubble diameters for nozzle, capillary and diffuser differ in (a) by $\approx 10\% - 30\%$. This is a fact, which is neglected in nowadays standard effective models and correlations, respectively, cf. [Jam01, Gad86, Voi87].

An interesting issue is given by the curve of capillary and diffuser in (a). For strong wetting and strong non–wetting behavior the two curves appear to result in the approximately same size of the diameter. An explanation for the strong wetting behavior is, that in both cases the base area of the orifice including the wetted fringe is comparable and

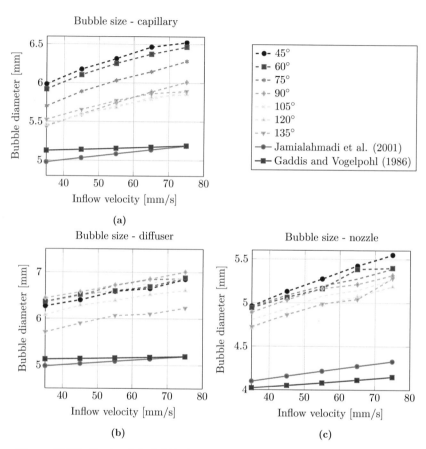

Figure (6.10): Results of the bubble diameter for different geometries under the influence of the inflow velocity and applied wetting properties for (a) a capillary inlet, (b) a diffuser inlet and (c) a nozzle inlet.

therefore the cross–section at the outlet of the orifice is comparable and bigger with respect to the cross–section at the inlet, giving rise to comparable big bubble sizes. An analogous consideration can be made for the non–wetting case. If the injected fluid is not wetting the surface and is only in contact with the interior wall of the orifice, the

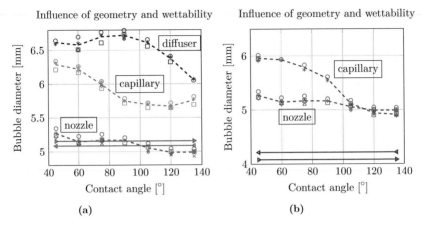

Figure (6.11): Influence of geometry and wettability on the bubble diameter for the same inflow volume flux. Comparison of the three different geometries for the same orifice inlet diameter of (a) $d = 3\,\mathrm{mm}$ with an inflow velocity of $v_{in} = 0.055\,\mathrm{m/s}$ and (b) the same orifice outlet diameter of $d = 1.4\,\mathrm{mm}$ with an outlet velocity at the orifice of $\approx 0.253\,\mathrm{m/s}$. Markers denote the "$\circ$" first, "$*$" second, "$\square$" third and the "$\diamond$" fourth bubble in the simulation. References of Jamialahmadi et al. [Jam01] marked by "$\triangleleft$" and Gaddis and Vogelpohl [Gad86] with "$\triangleright$".

cross–section for capillary and diffuser are again the same and less compared to the wetting case. Therefore smaller bubble sizes are expected, but similar for capillary and diffuser. In the transition area the partial wetting condition accompanied by the geometrical structure, in particular the stalling angle of the diffuser, play a severe role. This plot further reveals, that the results of the nozzle are less influenced by the wetting condition. A reason is, that the outlet velocity of the nozzle is increased by one order of magnitude compared to the capillary just because of the constriction. In this way surface tension is losing its influence over inertia. Fig. (b) supports the argumentation by comparable base areas for non–wetting regimes. The outlet diameters of capillary and nozzle are the same here. This means in the non-wetting regime only the interior wall is in contact with the injected fluid and the velocities, when leaving the orifices, are the same. Therefore, in non-wetting cases, the bubble diameters of capillary and nozzle are of comparable sizes. Further this plots reveals, that despite the higher inertia, the geometry of the orifice still has an influence in the wetting regime. In this way even with the same outlet diameters and velocities, the results of the diameter differ by about 10%, which may be caused by the broader accessible fringe in the capillary case, cf. Fig. 6.5(d) and Fig. 6.7(d).

Summing up the findings, the results of this section reveal, that in the absence of wetting, the shape of the orifice plays a minor role and that an outlet diameter of an orifice may be enough to characterize the system for given flow conditions. These circumstances can be achieved in two different ways. On the one hand the orifice type is losing its influence if the injected fluid is a highly non–wetting fluid and on the other hand the design of the orifice becomes neglectable when surface tension is losing its influence due to higher capillary or bond numbers. But if the injected fluid is a partial wetting fluid and for low bond and capillary numbers, orifice type and wettability play a crucial role and should be considered in the correlations as well.

Evaluation of the flow field

In order to understand the detachment process of a single bubble, it is not only important to know, what's happening right at the inlet, e.g. due to wetting, but also essential to know, how the flow field nearby looks like. All these effects act together and affect the bubble detachment and release process. This is why the flow field is now investigated as shown in Fig. 6.12. In the left half of (a) streamlines are plotted right in the moment before the thin throat snaps off. In the right half velocity vectors are used to visualize the flow field of this situation. The strength of the flow field in terms of the magnitude of fluid velocity is illustrated in the left half by the underlying color field from blue to red, where the upper limit of the color scale is adjusted in each plot to map the highest occurring velocity. In the right half, quantitative velocities are also illustrated by the color and furthermore by the length of each velocity vector. While the volume of the attached bubble increases in the bubble formation process, liquid is displaced by the injected gas. As a result, now a counter current evolves close to the two–phase interface of the bubble. This behavior is visualized in terms of the velocity vectors and the streamlines. The attached bubble assumes a round shape, which originates from an energetic consideration, because a system always tries to reach a lower energetic state. This represents an optimum due to the reduced flow resistance and due to the reduced thermodynamical potential, the free energy. In this way a circulation of liquid around a rising bubble occurs, giving rise to a constriction. The flow field of the counter current is pushing towards the thin throat, cf. the streamlines and the velocity vectors of (a). In this way the flow field is one contributor to the snap–off effect. The corresponding pressure distribution for this time step is shown in (b). In presence of gravity and an applied density ratio of both fluids, different hydrostatic pressure gradients emerge inside and outside the bubble, pushing the bubble towards the top. Hence, buoyancy is another contributor and together with the circulation around the bubble both effects support the release of the lighter bubble. Depending on the process conditions, other contributors may rise and the dominating terms may vary. The flow field is also responsible for the distortion of the bubble right after the snap–off and during the rise process, see (c). The corresponding pressure profile for this time step is shown in (d). While the first bubble rises, the formation of the second bubble already starts at the inlet. Further influences by the flow field and a general overview of the rising process in the bubble column is shown later.

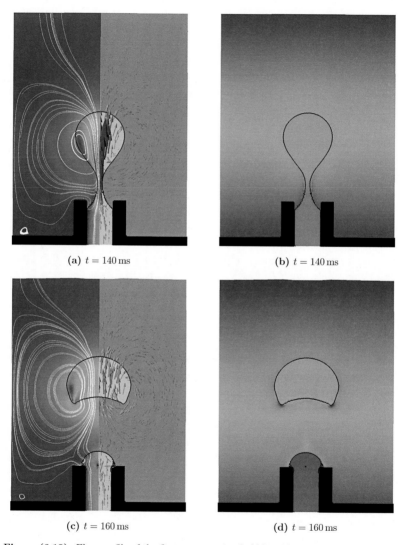

(a) $t = 140\,\mathrm{ms}$ **(b)** $t = 140\,\mathrm{ms}$

(c) $t = 160\,\mathrm{ms}$ **(d)** $t = 160\,\mathrm{ms}$

Figure (6.12): Flow profile of the first, non–wetting bubble with a contact angle of $135°$. (a) illustrates the bubble formation process before the detachment is accomplished with the corresponding qualitative pressure field shown in (b). In (c) the state after the detachment is shown with a rising bubble and the corresponding pressure profile in (d). On the left side of (a) and (c) streamlines of each time step are plotted and on the right side a random set of velocity vectors is used to illustrate the flow pattern.

To get an impression of the system under continuous conditions, now the detachment of
the second bubble is considered after the first one already contributed to the flow field. In
Fig. 6.13 the bubble detachment process of the capillary inlet for two different wetting
conditions is compared. In the left half of each picture, again streamlines are plotted
identifying the flow pattern in this time step for hypothetical massless elements and in the
right half velocity vectors are used to reveal the regions of high and low velocities. The
visualization of the wetting simulation (a) and the non–wetting simulation (c) show an
increased velocity in the constricted gaseous throat right before the snap–off happens. It is
observed, that the three–phase contact line is located in the latter case inside the capillary
at this time step, whereas in case of the wetting simulation the contact line sits right at the
outlet corner of the capillary. In the wetting case, the injected gaseous phase tries to cover
the surface of the capillary which is a contrary effect to the detachment process. Right
in the moment of the snap–off, a slight pullback of the two–phase interface is observed,
which is in case of the non–wetting gaseous phase stronger pronounced compared to the
wetting case due to the weaker tendency to wet the surface.

Apart from the position of the three–phase contact line and the different resulting bubble
volume, there is only a minor difference when comparing the wetting case to the non–
wetting case. Especially when the bubble is detached with respect to bubble shape and
the flow pattern in the free–flow area, see (b) and (d), the previous wetting influence is
negligible. The complete shape with the notch at the bottom of the bubble is due to
the comparable flow field by the circulation observed in both simulations. However, a
recognizable difference is obtained with regard to the residual gaseous phase, which stays
at the orifice after the snap–off. Here both simulations differ in the way they cover the
surface. The different base areas are again an indicator for the resulting bubble size as
explained in the previous section. (b) reveals, that for wetting conditions wider plateaus at
the end of capillaries lead to bigger bubbles. The velocity vectors show how the surface of
the capillary is wetted by the injected gaseous phase, which is contrary to the illustrated
behavior in (d). This leads to the assumption that in a non–wetting case a wider plateau
has no influence on the bubble size.

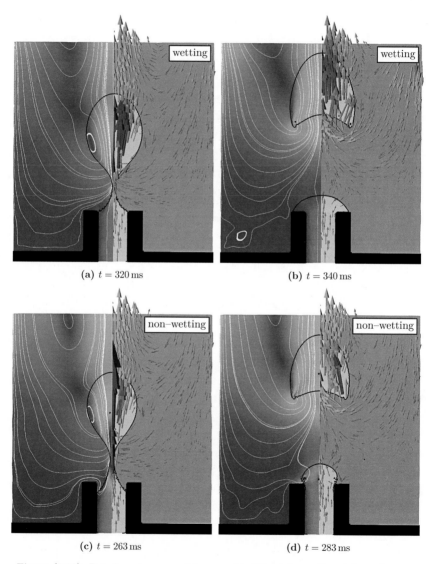

(a) $t = 320\,\text{ms}$ **(b)** $t = 340\,\text{ms}$

(c) $t = 263\,\text{ms}$ **(d)** $t = 283\,\text{ms}$

Figure (6.13): Detachment process of the second bubble at the capillary inlet with an inlet velocity of $0.075\,\text{m/s}$. (a) and (b) correspond to the wetting case with a contact angle of $45°$ for the injected fluid, (c) and (d) correspond to a non–wetting case with a contact angle of $135°$. The left part of the picture shows streamlines whereas on the right part a random set of velocity vectors of the flow field is shown.

Regarding the nozzle inlet in Fig. 6.14, the difference compared to the previous consideration with the capillary is, that for the same volume flux in the nozzle a higher outflow velocity is obtained at the narrower cross-section of the outlet. Looking at the wetting and the non–wetting case right before the snap-off, cf. Figs. (a) and (c), the two interface shapes are almost indistinguishable. The flow field as well as the covered base area are comparable, which gives rise to equal bubble sizes and thereby confirms in a 2D visualization the results plotted in the diagrams of Fig. 6.11. The geometrical shape of the nozzle leads to smoother streamlines especially in the bottom area of the column and in this way to a smoother circulation and flow field. These streamlines are tangentially directed with respect to the nozzles' surface and rather prevent a wetting of the gaseous phase on the surface. Moreover, this would mean, that the gas would have to creep down the inclination. In general due to buoyancy this is unlikely to happen, and only intermittently occurs in a less pronounced circulation right after a previous snap–off, cf. Figs. 6.14 (b) and (d). Last but not least the higher outflow velocity additionally reduces the influence of surface tension and increases the influence of inertia. In the moment of the snap–off, both wetting conditions result in comparable physical characteristics of both, flow field and covered base area. These circumstances are responsible for the minor sensitivity of the nozzle with respect to wetting conditions.

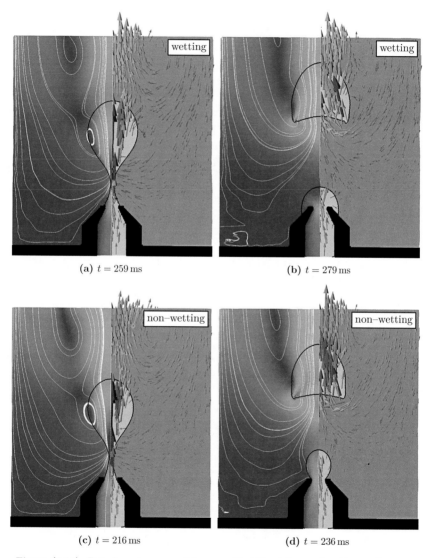

(a) $t = 259\,\text{ms}$ (b) $t = 279\,\text{ms}$

(c) $t = 216\,\text{ms}$ (d) $t = 236\,\text{ms}$

Figure (6.14): Detachment process of the second bubble at the nozzle with an inlet velocity of $0.075\,\text{m/s}$. (a) and (b) correspond to the wetting case with a contact angle of $45°$ for the injected fluid, (c) and (d) correspond to a non–wetting case with a contact angle of $135°$. The left part of the picture shows streamlines whereas on the right part a random set of velocity vectors of the flow field is shown.

In case of the diffuser, cf. Fig. 6.15, the same volume flux leads due to the opening angle to the lowest outflow velocity. This means inertia is again loosing influence and surface tension in combination with buoyancy become the dominating forces. If a wetting fluid is injected, first surface tension and buoyancy don't compete until the whole surface of the diffuser is wetted. Then, with increasing bubble volume and an emerging thinner and thinner bubble throat, buoyancy is further pulling on the bubble, whereas surface tension is trying to keep the contact line in its equilibrium position, cf. (a). As soon as buoyancy prevails, the snap–off happens, see (b). The bubble shape after the detachment process is again comparable to the previous simulations of capillary and nozzle. In the non–wetting case, cf. (c), the two–phase interface along with the contact line is retreated in the moment right before the snap–off, which is again comparable to the interface position of the capillary. Moreover, this analogy is the cause for the comparable bubble diameters of capillary and diffuser for injected non–wetting fluids, cf. Fig. 6.11(a). After the bubble is released, Fig. 6.15(d) shows that less surface of the diffuser is wetted, compared to the wetting case. This is an indication for resulting smaller bubble sizes. After the release the flow fields are again comparable among all orifice types and all investigated wetting conditions. The only difference is the resulting bubble volume and in combination with a constant inflow volume flux the detachment rate of the bubbles.

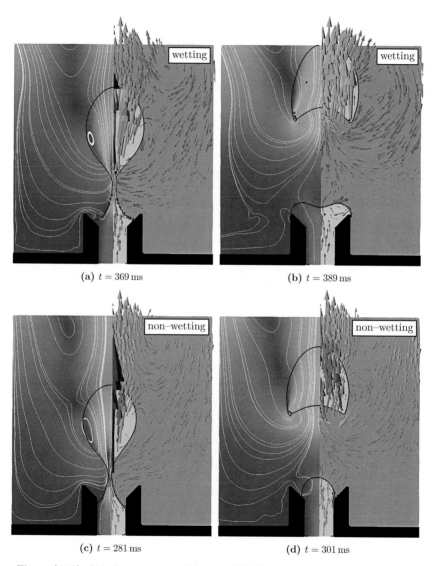

(a) $t = 369\,\text{ms}$ (b) $t = 389\,\text{ms}$

(c) $t = 281\,\text{ms}$ (d) $t = 301\,\text{ms}$

Figure (6.15): Detachment process of the second bubble at the diffuser with an inlet velocity of $0.075\,\text{m/s}$. (a) and (b) correspond to the wetting case with a contact angle of $45°$ for the injected fluid, (c) and (d) correspond to a non–wetting case with a contact angle of $135°$. The left part of the picture shows streamlines whereas on the right part a random set of velocity vectors of the flow field is shown.

A closer look at the dynamics of the contact line and the initial interface propagation in the bubble formation process is illustrated in Fig. 6.16. In the series of (a)–(d) a detailed plot of the capillary inlet is shown, with the first two plots belonging to the wetting case and the last two plots belonging to the non–wetting case. (a) and (c) reveal the different interface positions before the snap–off. The streamlines in the latter case also express the motion of the contact line to the retreated position. In (b) and (d) the different base areas are illustrated, where in the latter case the streamlines also emphasize how the non–wetting character of the injected fluid influences the flow field.

The series from (e)–(h) show for comparison the same situation for the nozzle. (e) and (g) visualize how similar the interface position and the flow fields are for the wetting and the non–wetting case in the moment right before the snap–off. This confirms the constant bubble diameter for various wetting conditions in case of the nozzle, cf. Fig. 6.11(a). Figs. 6.16(f) and 6.16(h) show the different wetted surfaces in the transient state between two snap–offs.

The series from (i)–(l) illustrate the corresponding time steps for the diffuser. (i) and (k) show the interface position for the wetting and the non–wetting case, where again a retreated interface is observed in the non–wetting case. (j) and (l) reveal the different base areas, which give a hint on different bubble volumes.

The influence by the external flow field is also observed in these plots, as the symmetry is lost in the simulations, which can already be caused by tiny disturbances like summing up round–off errors over time. The non–wetting plots of capillary and diffuser reveal the similarity of both simulations for injected non–wetting fluids and therefore give an explanation for the converging bubble diameter curves in Fig. 6.11(a). In the same way, due to a comparable contact line position, these simulations reveal comparable bubble diameters for capillary and diffuser in strong wetting regimes. Moreover, a discrepancy to the nozzle is observed in the wetting case. The bubble shape of the nozzle in case of the non–wetting fluid (h) is comparable to the capillary (d) and the diffuser (l), which confirms comparable bubble sizes for comparable outlet diameters, cf. Fig. 6.11(a).

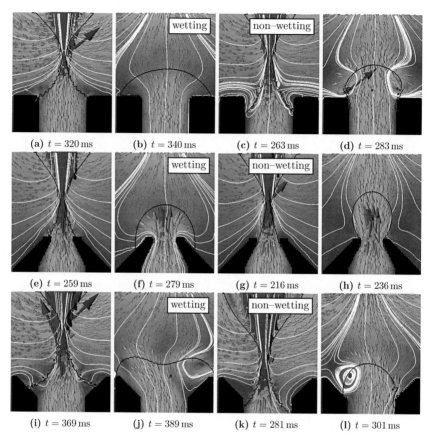

Figure (6.16): Detailed plot of the capillary inlet: Wetting fluid with $\alpha_S = 45°$ shown in (a) and (b). Non–wetting fluid with $\alpha_S = 135°$ shown in (c) and (d). Detailed plot of the nozzle inlet: Wetting fluid with $\alpha_S = 45°$ shown in (e) and (f). Non–wetting fluid with $\alpha_S = 135°$ shown in (g) and (h). Detailed plot of the diffuser inlet: Wetting fluid with $\alpha_S = 45°$ shown in (i) and (j). Non–wetting fluid with $\alpha_S = 135°$ shown in (k) and (l).

After the detailed analysis of the behavior at the inlet now the influence of the outer flow field is further investigated, see Fig. 6.17. The impact on the bubble detachment is illustrated in Fig. (a). The streamlines and velocity vectors reveal how the flow field couples the motion of the upper bubble with detachment of the lower bubble by its vortices. In this manner the flow field is in the same way a result of the bubble motion as well as a cause thereof. There is an influence of the flow pattern on the bubble shape as well as an impact of the bubble shape on the resulting flow pattern. This corresponds to classical two–way coupling if the bubbles would be described in a Lagrangian manner by furthermore neglecting direct bubble–bubble interactions. During the course of simulation the flow field further influences the evolution of the bubble shape, cf. Fig. (b). Since the snap–off of the lower bubble the velocity profile inside the lower bubble or in the bubble wake respectively is higher compared to the velocity outside the lower bubble. This is the reason why the second bubble almost reaches the first one, see Fig. (c) and (d). The slow liquid phase velocity above the upper bubble is previously causing a retardation of the rise process. This retardation is further enhanced by the stronger resistance due to the flat bubble shape, which leads to a pursuit of the lower bubble in the wake of the upper one.

In a two–dimensional and in a pseudo three–dimensional simulation, where an isotropic system by rotational invariance is assumed, capillary waves are rather suppressed in contrast to a general three–dimensional consideration. There, the additional dimension gives rise to a further degree of freedom, which can be utilized by an interface modulation. Therefore, the previously shown simulation results are less sensitive towards perturbations as compared to real experiments and complete three–dimensional simulations. Capillary waves are mainly observed while the bubble rise, because the interaction in motion with the outer fluid may act as excitation. Usually, when the bubble is formed at the orifice, capillary waves don't instantaneously show up. They are driven by small perturbations of the flow field close to the two–phase interface. In this way a stochastic complex behavior may affect interface dynamics and lead to an excitation of capillary waves. This characteristic behavior depends further on interface conditions such as surface tension, viscosity and density ratio etc..

The subject of this work is to solely focus on the influence of wettability and different orifice types on single bubble formation in order to emphasize this dependency without other, outer influences. A smart choice of the computational domain with appropriate boundary conditions can moreover be used to minimize other external influences. Hence, a rather regular flow field is obtained in these simulations, which allows for a really focussed investigation of primary bubble formation at column reactors.

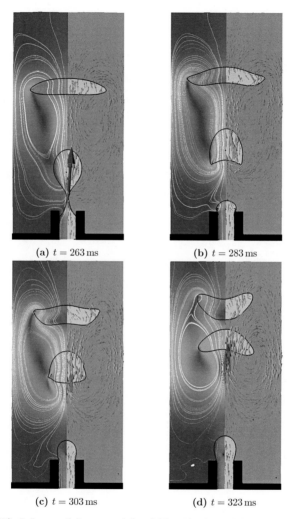

(a) $t = 263\,\text{ms}$ **(b)** $t = 283\,\text{ms}$

(c) $t = 303\,\text{ms}$ **(d)** $t = 323\,\text{ms}$

Figure (6.17): Influence of the external flow field on the detachment of the second bubble for the non–wetting case with a contact angle of $135°$ at the capillary and an inlet velocity of $0.075\,\text{m/s}$.

Bibliography

[Ada10] ADAMI, S.; HU, X. Y. und ADAMS, N. A.: A new surface-tension formulation for multi-phase SPH using a reproducing divergence approximation. *Journal of Computational Physics* (2010), Bd. 229(13): S. 5011–5021

[Ada12] ADAMI, S.; HU, X. Y. und ADAMS, N. A.: A generalized wall boundary condition for smoothed particle hydrodynamics. *Journal of Computational Physics* (2012), Bd. 231(21): S. 7057–7075

[Ada13] ADAMI, S.; HU, X.Y. und ADAMS, N.A.: A transport-velocity formulation for smoothed particle hydrodynamics. *Journal of Computational Physics* (2013), Bd. 241(0): S. 292–307

[AM11a] ACEVEDO-MALAVE, A. und GARCIA-SUCRE, M.: Coalescence collision of liquid drops I: Off-center collisions of equal-size drops. *Aip Advances* (2011), Bd. 1(3): S. 032117

[AM11b] ACEVEDO-MALAVE, A. und GARCIA-SUCRE, M.: Coalescence collision of liquid drops II: Off-center collisions of unequal-size drops. *Aip Advances* (2011), Bd. 1(3): S. 032118

[Ami00] AMIRFAZLI, A.; HANIG, S.; MULLER, A. und NEUMANN, A. W.: Measurements of line tension for solid-liquid-vapor systems using drop size dependence of contact angles and its correlation with solid-liquid interfacial tension. *Langmuir* (2000), Bd. 16(4): S. 2024–2031

[Ann05] ANNALAND, M. V.; DEEN, N. G. und KUIPERS, J. A. M.: Numerical simulation of gas bubbles behaviour using a three-dimensional volume of fluid method. *Chemical Engineering Science* (2005), Bd. 60(11): S. 2999–3011

[Bal97] BALAY, Satish; GROPP, William D.; MCINNES, Lois Curfman und SMITH, Barry F.: Efficient Management of Parallelism in Object Oriented Numerical Software Libraries, in: E. Arge; A. M. Bruaset und H. P. Langtangen (Herausgeber)

Modern Software Tools in Scientific Computing, Birkhäuser Press (1997), S. 163–202

[Bal13] BALAY, Satish; ABHYANKAR, Shrirang; ADAMS, Mark F.; BROWN, Jed; BRUNE, Peter; BUSCHELMAN, Kris; EIJKHOUT, Victor; GROPP, William D.; KAUSHIK, Dinesh; KNEPLEY, Matthew G.; McINNES, Lois Curfman; RUPP, Karl; SMITH, Barry F. und ZHANG, Hong: PETSc Users Manual, Techn. Ber. ANL-95/11 - Revision 3.4, Argonne National Laboratory (2013)

[Bal14] BALAY, Satish; ABHYANKAR, Shrirang; ADAMS, Mark F.; BROWN, Jed; BRUNE, Peter; BUSCHELMAN, Kris; EIJKHOUT, Victor; GROPP, William D.; KAUSHIK, Dinesh; KNEPLEY, Matthew G.; McINNES, Lois Curfman; RUPP, Karl; SMITH, Barry F. und ZHANG, Hong: PETSc Web page, http://www.mcs.anl.gov/petsc (2014)

[Ban13] BANDARA, U. C.; TARTAKOVSKY, A. M.; OOSTROM, M.; PALMER, B. J.; GRATE, J. und ZHANG, C.: Smoothed particle hydrodynamics pore-scale simulations of unstable immiscible flow in porous media. *Advances In Water Resources* (2013), Bd. 62: S. 356–369

[Bas09] BASA, M.; QUINLAN, N. J. und LASTIWKA, M.: Robustness and accuracy of SPH formulations for viscous flow. *International Journal For Numerical Methods In Fluids* (2009), Bd. 60(10): S. 1127–1148

[Bec94] BECKER, S.; SOKOLICHIN, A. und EIGENBERGER, G.: Gas–liquidiquid flow in bubble columns and loop reactors: Part II. Comparison of detailed experiments and flow simulations. *Chemical Engineering Science* (1994), Bd. 49(24, Part 2): S. 5747 – 5762

[Bel98] BELYTSCHKO, T.; KRONGAUZ, Y.; DOLBOW, J. und GERLACH, C.: On the completeness of meshfree particle methods. *International Journal For Numerical Methods In Engineering* (1998), Bd. 43(5): S. 785–819

[Bil11] BILOTTA, G.; RUSSO, G.; HERAULT, A. und DEL NEGRO, C.: Moving least-squares corrections for smoothed particle hydrodynamics. *Annals of Geophysics* (2011), Bd. 54(5): S. 622–633

[Bla69] BLAKE, T. D. und HAYNES, J. M.: Kinetics of Liquid/liquid Displacement. *Journal of Colloid and Interface Science* (1969), Bd. 30(3): S. 421–423

[Bla79] BLAKE, T. D. und RUSCHAK, K. J.: Maximum Speed of Wetting. *Nature* (1979), Bd. 282(5738): S. 489–491

[Bla88] BLASS, Eckhart: Bildung und Koaleszenz von Blasen und Tropfen. *Chemie Ingenieur Technik* (1988), Bd. 60(12): S. 935–947

[Bla93] BLAKE, T. D.: Dynamic contact angles and wetting kinetics, in: John C. Berg (Herausgeber) *Wettability*, Bd. 49 von *surfactant science series*, Kap. 5, M. Dekker (1993), S. 251–309

[Bla02] BLAKE, T. D. und SHIKHMURZAEV, Y. D.: Dynamic wetting by liquids of different viscosity. *Journal of Colloid and Interface Science* (2002), Bd. 253(1): S. 196–202

[Bon99] BONET, J. und LOK, T. S. L.: Variational and momentum preservation aspects of Smooth Particle Hydrodynamic formulations. *Computer Methods In Applied Mechanics and Engineering* (1999), Bd. 180(1-2): S. 97–115

[Bos93] BOSE, A.: Wetting by Solutions, in: John C. Berg (Herausgeber) *Wettability*, Bd. 49 von *surfactant science series*, Kap. 3, M. Dekker (1993), S. 149–181

[Bot04] BOTHE, D.; KOEBE, M.; WIELAGE, K.; PRUESS, J. und WARNECKE, H.-J.: Direct numerical simulation of mass transfer between rising gas bubbles and water, in: Martin Sommerfeld (Herausgeber) *Bubbly Flows*, Heat and Mass Transfer, Springer Berlin Heidelberg (2004), S. 159–174

[Bra92] BRACKBILL, J. U.; KOTHE, D. B. und ZEMACH, C.: A Continuum Method For Modeling Surface-tension. *Journal of Computational Physics* (1992), Bd. 100(2): S. 335–354

[Bre13] BREINLINGER, Thomas; POLFER, Pit; HASHIBON, Adham und KRAFT, Torsten: Surface Tension and Wetting Effects with Smoothed Particle Hydrodynamics. *Journal of Computational Physics* (2013), Bd. 243(0): S. 14–27

[Bro85] BROOKSHAW, L.: A Method of Calculating Radiative Heat Diffusion In Particle Simulations. *Proceedings Astronomical Society of Australia* (1985), Bd. 6(2): S. 207–210

[Bro92] BROCHARDWYART, F. und DEGENNES, P. G.: Dynamics of Partial Wetting. *Advances In Colloid and Interface Science* (1992), Bd. 39: S. 1–11

[Cas44] CASSIE, A. B. D. und BAXTER, S.: Wettability of porous surfaces. *Trans. Faraday Soc.* (1944), Bd. 40: S. 546–551

[Che99] CHEN, J. K.; BERAUN, J. E. und CARNEY, T. C.: A corrective smoothed particle method for boundary value problems in heat conduction. *International Journal For Numerical Methods In Engineering* (1999), Bd. 46(2): S. 231–252

[Che13] CHEN, Z.; ZONG, Z.; LIU, M. B. und LI, H. T.: A comparative study of truly incompressible and weakly compressible SPH methods for free surface incompressible flows. *International Journal For Numerical Methods In Fluids* (2013), Bd. 73(9): S. 813–829

[Che14] CHEN, Longquan und BONACCURSO, Elmar: Effects of surface wettability and liquid viscosity on the dynamic wetting of individual drops. *Phys. Rev. E* (2014), Bd. 90: S. 022401

[Cho68] CHORIN, Alexandre Joel: Numerical Solution of the Navier-Stokes Equations. *Mathematics of Computation* (1968), Bd. 22(104): S. 745–762

[Col03] COLAGROSSI, A. und LANDRINI, M.: Numerical simulation of interfacial flows by smoothed particle hydrodynamics. *Journal of Computational Physics* (2003), Bd. 191(2): S. 448–475

[Cum99] CUMMINS, Sharen J. und RUDMAN, Murray: An SPH Projection Method. *Journal of Computational Physics* (1999), Bd. 152(2): S. 584–607

[Das09] DAS, A. K. und DAS, P. K.: Bubble evolution through submerged orifice using smoothed particle hydrodynamics: Basic formulation and model validation. *Chemical Engineering Science* (2009), Bd. 64(10): S. 2281–2290

[Das10] DAS, A.K. und DAS, P.K.: Equilibrium shape and contact angle of sessile drops of different volumes–Computation by SPH and its further improvement by DI. *Chemical Engineering Science* (2010), Bd. 65(13): S. 4027–4037

[Das11] DAS, A. K. und DAS, P. K.: Incorporation of diffuse interface in smoothed particle hydrodynamics: Implementation of the scheme and case studies. *International Journal For Numerical Methods In Fluids* (2011), Bd. 67(6): S. 671–699

[Dee01] DEEN, N. G.; SOLBERG, T. und HJERTAGER, B. H.: Large eddy simulation of the gas-liquid flow in a square cross-sectioned bubble column. *Chemical Engineering Science* (2001), Bd. 56(21-22): S. 6341–6349 [deG85] DEGENNES, P. G.: Wetting - Statics and Dynamics. *Reviews of Modern Physics* (1985), Bd. 57(3): S. 827–863

[deG03] DEGENNES, P. G.; BROCHARDWYART, F. und QUERE, D.: *Capillarity and Wetting Phenomena: Drops, Bubbles, Pearls, Waves*, Springer (2003)

[Dob15] DOBESCH, Daniel: Direkte numerische Simulation von Einspritzvorgängen mit Tropfenbildung, Bachelorarbeit, Universität Stuttgart (2015)

[Dwe12] DWENGER, S.; EIGENBERGER, G. und NIEKEN, U.: Measurement of Capillary Pressure-Saturation Relationships Under Defined Compression Levels for Gas Diffusion Media of PEM Fuel Cells. *Transport In Porous Media* (2012), Bd. 91(1): S. 281–294

[Dwe15] DWENGER, Stefan: *Transport interactions between gas and water in thin hydrophobic porous layers*, Dissertation, University of Stuttgart (2015)

[Fal02] FALGOUT, Robert D. und YANG, Ulrike Meier: hypre: a Library of High Performance Preconditioners, in: *Preconditioners,"* Lecture Notes in Computer Science, S. 632–641

[Fal06] FALGOUT, RobertD.; JONES, JimE. und YANG, UlrikeMeier: The Design and Implementation of hypre, a Library of Parallel High Performance Preconditioners, in: AreMagnus Bruaset und Aslak Tveito (Herausgeber) *Lecture Notes in Computational Science and Engineering*, Bd. 51, Springer Berlin Heidelberg (2006), S. 267–294

[Fan06] FANG, J. N.; OWENS, R. G.; TACHER, L. und PARRIAUX, A.: A numerical study of the SPH method for simulating transient viscoelastic free surface flows. *Journal of Non-newtonian Fluid Mechanics* (2006), Bd. 139(1-2): S. 68–84

[Fat11] FATEHI, R. und MANZARI, M. T.: Error estimation in smoothed particle hydrodynamics and a new scheme for second derivatives. *Computers & Mathematics With Applications* (2011), Bd. 61(2): S. 482–498

[Fat14] FATEHI, R.; SHADLOO, M. S. und MANZARI, M. T.: Numerical investigation of twophase secondary Kelvin-Helmholtz instability. *Proceedings of the Institution of Mechanical Engineers Part C-journal of Mechanical Engineering Science* (2014), Bd. 228(11): S. 1913–1924

[Fer13a] FERRAND, M.; LAURENCE, D. R.; ROGERS, B. D.; VIOLEAU, D. und KASSIOTIS,

C.: Unified semi-analytical wall boundary conditions for inviscid, laminar or turbulent flows in the meshless SPH method. *International Journal For Numerical Methods In Fluids* (2013), Bd. 71(4): S. 446–472

[Fer13b] FERRARI, Andrea und LUNATI, Ivan: Direct numerical simulations of interface dynamics to link capillary pressure and total surface energy. *Advances in Water Resources* (2013), Bd. 57(0): S. 19–31

[Fle96] FLEISCHER, C.; BECKER, S. und EIGENBERGER, G.: Detailed modeling of the chemisorption of CO2 into NaOH in a bubble column. *Chemical Engineering Science* (1996), Bd. 51(10): S. 1715–1724

[Gad86] GADDIS, E. S. und VOGELPOHL, A.: Bubble Formation In Quiescent Liquids Under Constant Flow Conditions. *Chemical Engineering Science* (1986), Bd. 41(1): S. 97–105

[GG12] GOMEZ-GESTEIRA, M.; ROGERS, B. D.; CRESPO, A. J. C.; DALRYMPLE, R. A.; NARAYANASWAMY, M. und DOMINGUEZ, J. M.: SPHysics - development of a free-surface fluid solver - Part 1: Theory and formulations. *Computers & Geosciences* (2012), Bd. 48: S. 289–299

[Gin77] GINGOLD, R. A. und MONAGHAN, J. J.: Smoothed Particle Hydrodynamics - Theory And Application To Non-Spherical Stars. *Monthly Notices Of The Royal Astronomical Society* (1977), Bd. 181(2): S. 375–389

[Gin82] GINGOLD, R. A. und MONAGHAN, J. J.: Kernel Estimates As A Basis For General Particle Methods In Hydrodynamics. *Journal Of Computational Physics* (1982), Bd. 46(3): S. 429–453

[Gov57] GOVIER, G. W.; RADFORD, B. A. und DUNN, J. S. C.: The upward vertical flow of air-water mixtures: I. Effect of air and water rates on flow pattern, holdup and pressure drop. *The Canadian Journal of Chemical Engineering* (1957), Bd. 35: S. 58–70

[Gov58] GOVIER, G. W. und SHORT, W. Leigh: The upward vertical flow of air-water mixtures: II. Effect of tubing diameter on flow-pattern, holdup and pressure drop. *The Canadian Journal of Chemical Engineering* (1958), Bd. 36(5): S. 195–202

[Gre09] GRENIER, N.; ANTUONO, M.; COLAGROSSI, A.; LE TOUZE, D. und ALESSANDRINI,

B.: An Hamiltonian interface SPH formulation for multi-fluid and free surface flows. *Journal of Computational Physics* (2009), Bd. 228(22): S. 8380–8393

[Gre13] GRENIER, N.; LE TOUZE, D.; COLAGROSSI, A.; ANTUONO, M. und COLICCHIO, G.: Viscous bubbly flows simulation with an interface SPH model. *Ocean Engineering* (2013), Bd. 69(0): S. 88–102

[Häg15] HÄGELE, Vanessa: Validierung von Ein- und Mehrphasenströmungen für das korrigierte Smoothed Particle Hydrodynamics, Semesterarbeit, Universität Stuttgart (2015)

[Has93] HASSANIZADEH, S. M. und GRAY, W. G.: Thermodynamic Basis of Capillary-pressure In Porous-media. *Water Resources Research* (1993), Bd. 29(10): S. 3389–3405

[Hen13] HENKEL, Klaus-Dieter: Reactor Types and Their Industrial Applications, in: *Ullmann's Encyclopedia of Industrial Chemistry*, Bd. 31, Wiley-VCH Verlag GmbH & Co. KGaA (2013), S. 293–328

[Hir16] HIRSCHLER, Manuel; KUNZ, Philip; HUBER, Manuel; HAHN, Friedemann und NIEKEN, Ulrich: Open boundary conditions for ISPH and their application to micro-flow. *Journal of Computational Physics* (2016), Bd. 307: S. 614 – 633

[Hof75] HOFFMAN, Richard L: A study of the advancing interface. I. Interface shape in liquid—gas systems. *Journal of Colloid and Interface Science* (1975), Bd. 50(2): S. 228–241

[Hu06] HU, X.Y. und ADAMS, N.A.: A multi-phase SPH method for macroscopic and mesoscopic flows. *Journal of Computational Physics* (2006), Bd. 213(2): S. 844–861

[Hu07] HU, X.Y. und ADAMS, N.A.: An incompressible multi-phase SPH method. *Journal of Computational Physics* (2007), Bd. 227(1): S. 264–278

[Hub13] HUBER, M.; SÄCKEL, W.; HIRSCHLER, M.; HASSANIZADEH, S.M. und NIEKEN, U.: Modeling the dynamics of partial wetting, in: *Particles 2013*, III International Conference on Particle-based Methods – Fundamentals and Applications

[Hub16a] HUBER, M.; DOBESCH, D.; KUNZ, P.; HIRSCHLER, M. und NIEKEN, U.: Influence of orifice type and wetting properties on bubble formation at bubble column reactors. *Chemical Engineering Science* (2016), Bd. 152: S. 151 – 162

[Hub16b] HUBER, M.; KELLER, F.; SÄCKEL, W.; HIRSCHLER, M.; KUNZ, P.; HAS-
SANIZADEH, S.M. und NIEKEN, U.: On the physically based modeling of
surface tension and moving contact lines with dynamic contact angles on the
continuum scale. *Journal of Computational Physics* (2016), Bd. 310: S. 459–477

[Inu02] INUTSUKA, S.: Reformulation of smoothed particle hydrodynamics with Riemann
solver. *Journal of Computational Physics* (2002), Bd. 179(1): S. 238–267

[Ish11] ISHII, M. und HIBIKI, T.: *Thermo-Fluid Dynamics of Two-Phase Flow*, Springer-
Verlag New York (2011)

[Jac00] JACQMIN, D.: Contact-line dynamics of a diffuse fluid interface. *Journal of Fluid
Mechanics* (2000), Bd. 402: S. 57–88

[Jam01] JAMIALAHMADI, M.; ZEHTABAN, M.R.; MUELLER-STEINHAGEN, H.; SARRAFI, A.
und SMITH, J.M.: Study of Bubble Formation Under Constant Flow Conditions.
Chemical Engineering Research and Design (2001), Bd. 79(5): S. 523 – 532,
fluid Flow

[Jia10] JIANG, T.; JIE, O. Y.; YANG, B. X. und REN, J. L.: The SPH method for
simulating a viscoelastic drop impact and spreading on an inclined plate.
Computational Mechanics (2010), Bd. 45(6): S. 573–583

[Joh64] JOHNSON, R. E., R. E. und DETTRE, R. H., R. H.: Contact Angle Hysteresis .3.
Study of An Idealized Heterogeneous Surface. *Journal of Physical Chemistry*
(1964), Bd. 68(7): S. 1744–&

[Joh93] JOHNSON, R. E. und DETTRE, R. H.: Wetting of Low-Energy Surfaces, in: John C.
Berg (Herausgeber) *Wettability*, Bd. 49 von *surfactant science series*, Kap. 3,
M. Dekker (1993), S. 1–73

[Kel14] KELLER, Franz: *Simulation of the Morphogenesis of Open-porous Materials*,
Dissertation, University of Stuttgart (2014)

[Kim15] KIM, J. H. und ROTHSTEIN, J. P.: Dynamic contact angle measurements of
viscoelastic fluids. *Journal of Non-newtonian Fluid Mechanics* (2015), Bd.
225: S. 54–61

[Klu83] KLUG, P. und VOGELPOHL, A.: Calculation of Stable Secondary Bubble-size
In Gas-distribution On Perforated Plates. *Chemie Ingenieur Technik* (1983),
Bd. 55(10): S. 804–805

[Koy04] KOYNOV, A. und KHINAST, J. G.: Effects of hydrodynamics and Lagrangian transport on chemically reacting bubble flows. *Chemical Engineering Science* (2004), Bd. 59(18): S. 3907–3927

[Koy06] KOYNOV, A. A. und KHINAST, J. G.: Micromixing in reactive, deformable bubble, and droplet swarms. *Chemical Engineering & Technology* (2006), Bd. 29(1): S. 13–23

[Kun15] KUNZ, P.; ZARIKOS, I.M.; KARADIMITRIOU, N.K.; HUBER, M.; NIEKEN, U. und HASSANIZADEH, S.M.: Study of Multi-phase Flow in Porous Media: Comparison of SPH Simulations with Micro-model Experiments. *Transport in Porous Media* (2015): S. 1–20

[Kun16] KUNZ, P.; HIRSCHLER, M.; HUBER, M. und NIEKEN, U.: Inflow/outflow with Dirichlet boundary conditions for pressure in {ISPH}. *Journal of Computational Physics* (2016), Bd. 326: S. 171 – 187

[Laf94] LAFAURIE, B.; NARDONE, C.; SCARDOVELLI, R.; ZALESKI, S. und ZANETTI, G.: Modeling Merging and Fragmentation In Multiphase Flows With Surfer. *Journal of Computational Physics* (1994), Bd. 113(1): S. 134–147

[Lam02] LAM, C. N. C.; WU, R.; LI, D.; HAIR, M. L. und NEUMANN, A. W.: Study of the advancing and receding contact angles: liquid sorption as a cause of contact angle hysteresis. *Advances In Colloid and Interface Science* (2002), Bd. 96(1-3): S. 169–191

[Lan87] LANDAU, L. D. und LIFSHITZ, E. M.: *Fluid Mechanics*, Bd. 6, Pergamon Press, Headington Hill Hall, Oxford, England, second Aufl. (1987)

[Lee08] LEE, E. S.; MOULINEC, C.; XU, R.; VIOLEAU, D.; LAURENCE, D. und STANSBY, P.: Comparisons of weakly compressible and truly incompressible algorithms for the SPH mesh free particle method. *Journal of Computational Physics* (2008), Bd. 227(18): S. 8417–8436

[Li96] LI, S. F. und LIU, W. K.: Moving least-square reproducing kernel method .2. Fourier analysis. *Computer Methods In Applied Mechanics and Engineering* (1996), Bd. 139(1-4): S. 159–193

[Liu97] LIU, W. K.; LI, S. F. und BELYTSCHKO, T.: Moving least-square reproducing kernel methods .1. Methodology and convergence. *Computer Methods In Applied Mechanics and Engineering* (1997), Bd. 143(1-2): S. 113–154

[Liu03] LIU, G. R. und LIU, M. B.: *Smoothed Particle Hydrodynamics*, World Scientific
         Publ Co Pte Ltd (2003)

[Liu05] LIU, Jie; KOSHIZUKA, Seiichi und OKA, Yoshiaki: A hybrid particle-mesh method
         for viscous, incompressible, multiphase flows. *Journal of Computational Physics*
         (2005), Bd. 202(1): S. 65–93

[Luc77] LUCY, L. B.: Numerical Approach To Testing Of Fission Hypothesis. *Astronomical
         Journal* (1977), Bd. 82(12): S. 1013–1024

[Ma12]   MA, Dou; LIU, Mingyan; ZU, Yonggui und TANG, Can: Two-dimensional volume
         of fluid simulation studies on single bubble formation and dynamics in bubble
         columns. *Chemical Engineering Science* (2012), Bd. 72: S. 61–77

[Mah15] MAHADY, Kyle; AFKHAMI, Shahriar und KONDIC, Lou: A volume of fluid method
         for simulating fluid/fluid interfaces in contact with solid boundaries. *Journal
         of Computational Physics* (2015), Bd. 294: S. 243–257

[Mar71] MARANGONI, C.: Über die Ausbreitung der Tropfen einer Flüssigkeit auf der
         Oberfläche einer anderen. *Annalen der Physik und Chemie* (1871), Bd. 219: S.
         337–354

[Mas14] MASOOD, Rao M. A. und DELGADO, Antonio: Numerical Investigation of Three-
         Dimensional Bubble Column Flows: A Detached Eddy Simulation Approach.
         Chem. Eng. Technol. (2014), Bd. 37(10): S. 1697–1704

[May13] MAYRHOFER, A.; ROGERS, B. D.; VIOLEAU, D. und FERRAND, M.: Investigation
         of wall bounded flows using SPH and the unified semi-analytical wall boundary
         conditions. *Computer Physics Communications* (2013), Bd. 184(11): S. 2515–
         2527

[Mer77] MERSMANN, Alfons: Auslegung und Maßstabsvergrößerung von Blasen- und
         Tropfensäulen. *Chemie Ingenieur Technik* (1977), Bd. 49(9): S. 679–691

[Mon82] MONAGHAN, J. J.: Why Particle Methods Work. *Siam Journal On Scientific
         and Statistical Computing* (1982), Bd. 3(4): S. 422–433

[Mon85] MONAGHAN, J. J.: Particle Methods For Hydrodynamics. *Computer Physics
         Reports* (1985), Bd. 3(2): S. 71–124

[Mon92] MONAGHAN, J. J.: Smoothed Particle Hydrodynamics. *Annual Review Of As-
         tronomy And Astrophysics* (1992), Bd. 30: S. 543–574

[Mon05] MONAGHAN, J. J.: Smoothed particle hydrodynamics. *Reports On Progress In Physics* (2005), Bd. 68(8): S. 1703–1759

[Mon12] MONAGHAN, J. J.: Smoothed Particle Hydrodynamics and Its Diverse Applications (2012)

[Mon13] MONAGHAN, J. J. und RAFIEE, A.: A simple SPH algorithm for multi-fluid flow with high density ratios. *International Journal For Numerical Methods In Fluids* (2013), Bd. 71(5): S. 537–561

[Mor97] MORRIS, Joseph P.; FOX, Patrick J. und ZHU, Yi: Modeling Low Reynolds Number Incompressible Flows Using SPH. *Journal of Computational Physics* (1997), Bd. 136(1): S. 214–226

[Mor00] MORRIS, J. P.: Simulating surface tension with smoothed particle hydrodynamics. *International Journal For Numerical Methods In Fluids* (2000), Bd. 33(3): S. 333–353

[Osi62] OSIPOW, L. I.: *Surface Chemistry: Theory and Industrial Applications*, Nr. 153 in American Chemical Society Monograph Series, Reinhold Publishing Corporation (1962)

[Pan13] PAN, W.; TARTAKOVSKY, A. M. und MONAGHAN, J. J.: Smoothed particle hydrodynamics non-Newtonian model for ice-sheet and ice-shelf dynamics. *Journal of Computational Physics* (2013), Bd. 242: S. 828–842

[Pfl01] PFLEGER, D. und BECKER, S.: Modelling and simulation of the dynamic flow behaviour in a bubble column. *Chemical Engineering Science* (2001), Bd. 56(4): S. 1737 – 1747, 16th International Conference on Chemical Reactor Engineering

[Pin08] PINDER, G. F. und GRAY, W. G.: *Essentials of multiphase flow and transport in porous media*, Wiley (2008)

[Pri08] PRICE, D. J.: Modelling discontinuities and Kelvin-Helmholtz instabilities in SPH. *Journal of Computational Physics* (2008), Bd. 227(24): S. 10040–10057

[Qui06] QUINLAN, N. J.; BASA, M. und LASTIWKA, M.: Truncation error in mesh-free particle methods. *International Journal For Numerical Methods In Engineering* (2006), Bd. 66(13): S. 2064–2085

[RA12] RAVI ANNAPRAGADA, S.; MURTHY, Jayathi Y. und GARIMELLA, Suresh V.:

Droplet retention on an incline. *International Journal of Heat and Mass Transfer* (2012), Bd. 55(5–6): S. 1457–1465

[Rad08] RADL, S.; KOYNOV, A.; TRYGGVASON, G. und KHINAST, J. G.: DNS-based prediction of the selectivity of fast multiphase reactions: Hydrogenation of nitroarenes. *Chemical Engineering Science* (2008), Bd. 63(12): S. 3279–3291

[Rad10] RADL, S.; SUZZI, D. und KHINAST, J. G.: Fast Reactions in Bubbly Flows: Film Model and Micromixing Effects. *Industrial & Engineering Chemistry Research* (2010), Bd. 49(21): S. 10715–10729

[Rae81] RAEBIGER, N. und VOGELPOHL, A.: Bubble Formation In Stationary and In Moving Newtonian Liquids. *Chemie Ingenieur Technik* (1981), Bd. 53(12): S. 976–977

[Raf12] RAFIEE, Ashkan; CUMMINS, Sharen; RUDMAN, Murray und THIAGARAJAN, Krish: Comparative study on the accuracy and stability of SPH schemes in simulating energetic free-surface flows. *European Journal of Mechanics - B/Fluids* (2012), Bd. 36(0): S. 1–16

[Ran96] RANDLES, P. W. und LIBERSKY, L. D.: Smoothed particle hydrodynamics: Some recent improvements and applications. *Computer Methods In Applied Mechanics and Engineering* (1996), Bd. 139(1-4): S. 375–408

[Ray79] RAYLEIGH, Lord: On the Capillary Phenomena of Jets. *Proceedings of the Royal Society of London* (1879), Bd. 29(196-199): S. 71–97

[Rea10] READ, J. I.; HAYFIELD, T. und AGERTZ, O.: Resolving mixing in smoothed particle hydrodynamics. *Monthly Notices of the Royal Astronomical Society* (2010), Bd. 405(3): S. 1513–1530

[Rei07] REIS, T. und PHILLIPS, T. N.: Lattice Boltzmann model for simulating immiscible two-phase flows. *Journal of Physics A-mathematical and Theoretical* (2007), Bd. 40(14): S. 4033–4053

[Ren07] REN, W. Q. und E, W. N.: Boundary conditions for the moving contact line problem. *Physics of Fluids* (2007), Bd. 19(2): S. 022101

[Ren11] REN, W. Q. und WEINAN, E.: Contact line dynamics on heterogeneous surfaces. *Physics of Fluids* (2011), Bd. 23(7): S. 072103

[Ros09] ROSSWOG, S.: Astrophysical smooth particle hydrodynamics. *New Astronomy Reviews* (2009), Bd. 53(4-6): S. 78–104

[Sca99] SCARDOVELLI, R. und ZALESKI, S.: Direct numerical simulation of free-surface and interfacial flow. *Annual Review of Fluid Mechanics* (1999), Bd. 31: S. 567–603

[Sch85] SCHWARTZ, L. W. und GAROFF, S.: Contact-angle Hysteresis On Heterogeneous Surfaces. *Langmuir* (1985), Bd. 1(2): S. 219–230

[Sef15] SEFID, M.; FATEHI, R. und SHAMSODDINI, R.: A Modified Smoothed Particle Hydrodynamics Scheme to Model the Stationary and Moving Boundary Problems for Newtonian Fluid Flows. *Journal of Fluids Engineering-transactions of the Asme* (2015), Bd. 137(3): S. 031201

[Sha82] SHAH, Y. T.; KELKAR, B. G.; GODBOLE, S. P. und DECKWER, W.-D.: Design parameters estimations for bubble column reactors. *AIChE Journal* (1982), Bd. 28(3): S. 353–379

[Sha03] SHAO, Songdong und LO, Edmond Y. M.: Incompressible SPH method for simulating Newtonian and non-Newtonian flows with a free surface. *Advances in Water Resources* (2003), Bd. 26(7): S. 787–800

[She68] SHEPARD, Donald: A two-dimensional interpolation function for irregularly-spaced data, in: *Proceedings of the 1968 23rd ACM national conference*, ACM '68, ACM, New York, NY, USA, S. 517–524

[Shi94] SHIKHMURZAEV, Y. D.: Mathematical-modeling of Wetting Hydrodynamics. *Fluid Dynamics Research* (1994), Bd. 13(1): S. 45–64

[Shi96] SHIKHMURZAEV, Y. D.: Dynamic contact angles and flow in vicinity of moving contact line. *Aiche Journal* (1996), Bd. 42(3): S. 601–612

[Shi97] SHIKHMURZAEV, Y. D.: Moving contact lines in liquid/liquid/solid systems. *Journal of Fluid Mechanics* (1997), Bd. 334: S. 211–249

[Sik05] SIKALO, S.; WILHELM, H. D.; ROISMAN, I. V.; JAKIRLIC, S. und TROPEA, C.: Dynamic contact angle of spreading droplets: Experiments and simulations. *Physics of Fluids* (2005), Bd. 17(6): S. 062103

[Sir13] SIROTKIN, F. V. und YOH, J. J.: A Smoothed Particle Hydrodynamics method with approximate Riemann solvers for simulation of strong explosions. *Computers & Fluids* (2013), Bd. 88: S. 418–429

[Sok94] SOKOLICHIN, A. und EIGENBERGER, G.: Gas–liquid flow in bubble columns and loop reactors: Part I. Detailed modelling and numerical simulation. *Chemical Engineering Science* (1994), Bd. 49(24, Part 2): S. 5735 – 5746

[Sok04] SOKOLICHIN, A.; EIGENBERGER, G. und LAPIN, A.: Simulation of buoyancy driven bubbly flow: Established simplifications and open questions. *AIChE Journal* (2004), Bd. 50(1): S. 24–45

[Sze12a] SZEWC, K.; POZORSKI, J. und MINIER, J. P.: Analysis of the incompressibility constraint in the smoothed particle hydrodynamics method. *International Journal For Numerical Methods In Engineering* (2012), Bd. 92(4): S. 343–369

[Sze12b] SZEWC, K.; TANIERE, A.; POZORSKI, J. und MINIER, J. P.: A Study on Application of Smoothed Particle Hydrodynamics to Multi-Phase Flows. *International Journal of Nonlinear Sciences and Numerical Simulation* (2012), Bd. 13(6): S. 383–395

[Sze15] SZEWC, K.; POZORSKI, J. und MINIER, J. P.: Spurious interface fragmentation in multiphase SPH. *International Journal For Numerical Methods In Engineering* (2015), Bd. 103(9): S. 625–649

[Tar05] TARTAKOVSKY, A. und MEAKIN, P.: Modeling of surface tension and contact angles with smoothed particle hydrodynamics. *Physical Review E* (2005), Bd. 72(2): S. 026301

[Tar15] TARTAKOVSKY, A.M.; TRASK, N.; PAN, K.; JONES, B.; PAN, W. und WILLIAMS, J.R.: Smoothed particle hydrodynamics and its applications for multiphase flow and reactive transport in porous media. *Computational Geosciences* (2015): S. 1–28

[Try11] TRYGGVASON, G.; SCARDOVELLI, R. und ZALESKI, S.: *Direct Numerical Simulations of Gas–Liquid Multiphase Flows*, Cambridge University Press (2011)

[Vac12] VACONDIO, R.; ROGERS, B. D. und STANSBY, P. K.: Smoothed Particle Hydrodynamics: Approximate zero-consistent 2-D boundary conditions and still shallow-water tests. *International Journal For Numerical Methods In Fluids* (2012), Bd. 69(1): S. 226–253

[Val10] VALCKE, S.; DE RIJCKE, S.; RODIGER, E. und DEJONGHE, H.: Kelvin-Helmholtz instabilities in smoothed particle hydrodynamics. *Monthly Notices of the Royal Astronomical Society* (2010), Bd. 408(1): S. 71–86

[Ver06] VEREIN DEUTSCHER INGENIEURE VDI-GESELLSCHAFT VERFAHRENSTECHNIK UND CHEMIEINGENIEURWESEN (GVC) (Herausgeber): *VDI-Wärmeatlas*, Springer Berlin Heidelberg (2006)

[Voi87] VOIT, Harald; ZEPPENFELD, Rolf und MERSMANN, Alfons: Calculation of primary bubble volume in gravitational and centrifugal fields. *Chemical Engineering & Technology* (1987), Bd. 10(1): S. 99–103

[Wen36] WENZEL, Robert N.: Resistance of solid surfaces to wetting by water. *Industrial & Engineering Chemistry* (1936), Bd. 28(8): S. 988–994

[Wen95] WENDLAND, Holger: Piecewise polynomial, positive definite and compactly supported radial functions of minimal degree. *Advances in Computational Mathematics* (1995), Bd. 4(1): S. 389–396

[Wil92] WILKINSON, Peter M.; SPEK, Arie P. und VAN DIERENDONCK, Laurent L.: Design parameters estimation for scale-up of high-pressure bubble columns. *AIChE Journal* (1992), Bd. 38(4): S. 544–554

[Woo11] WOOG, T.: *Partikelbasierte Simulationen zur Fluiddynamik unter Betrachtung von freien Oberflächen und Oberflächenspannung mit SPH*, Diplomarbeit, Universität Stuttgart (2011)

[Yan07] YANG, G. Q.; DU, B. und FAN, L. S.: Bubble formation and dynamics in gas-liquid-solid fluidization- A review. *Chemical Engineering Science* (2007), Bd. 62(1-2): S. 2–27

[Yan14] YANG, X. F.; LIU, M. B. und PENG, S. L.: Smoothed particle hydrodynamics modeling of viscous liquid drop without tensile instability. *Computers & Fluids* (2014), Bd. 92: S. 199–208

[Yuj12] YUJIE, Zhang; MINGYAN, Liu; YONGGUI, Xu und CAN, Tang: Three-dimensional volume of fluid simulations on bubble formation and dynamics in bubble columns. *Chemical Engineering Science* (2012), Bd. 73: S. 55–78

[Zha04] ZHANG, G. M. und BATRA, R. C.: Modified smoothed particle hydrodynamics method and its application to transient problems. *Computational Mechanics* (2004), Bd. 34(2): S. 137–146

[Zho08] ZHOU, G. Z.; GE, W. und LI, J. H.: A revised surface tension model for macro-scale particle methods. *Powder Technology* (2008), Bd. 183(1): S. 21–26

List of Figures

List of Tables

157